Books by the author

Equine

The Natural Horse: Lessons from the Wild (1992, 2020)

Horse Owners Guide to Natural Hoof Care (1999)

Founder – Prevention and Cure the Natural Way (2001)

Guide To Booting Horses for Hoof Care Professionals (2002)

Paddock Paradise: A Guide to Natural Horse Boarding (2005)

The Natural Trim: Principles and Practice (2012)

The Healing Angle: Nature's Gateway to the Healing Field (2014)

Laminitis: An Equine Plague of Unconscionable Proportions (2016)

Training Manual: ISNHCP Natural Trim Training Program (2017)

the Hoof Balancer: A Unique Tool for Balancing Equine Hooves (2019)

The Natural Trim: Advanced Guidelines (2019)

The Natural Trim: Basic Guidelines (2019, 2022)

*Navicular Syndrome: Healing And Prevention Using the Principles and
        Practices of Natural Horse Care (2021)*

A Closer Approximation of ☉ *On the MATW Using An Infrared
        Thermometer With Laser Pointer Gun (2021)*

Other

The Canvas Tipi (1982)

*Guard Your Teeth: Why the Dental Industry Fails Us
        A Guide to Natural Dental Care* (2018, 2022)

Buckskin Tanner: A Guide to Natural Hide Tanning (2019)

*Cheyenne Tipi Notes (1903): Technical Insights Into 19th Century
        Plains Indian Bison Hide Tanning* (2019)

*Living Behind the Facade: Memoirs Of A Gay Man's Journey Through
        the 20th Century* (2019) George Somers with Jaime Jackson

Platform: A Humanitarian Model For An Egalitarian Society (2019)

Zoo Paradise: A New Model for Humane Zoological Gardens (2019)

Forthcoming

Horse Trek – Into the Mystic

Guard Your Teeth!

Why the Dental Industry Fails Us

A Guide to
Natural Dental Care

Jaime Jackson

Natural World Publications

©2019, 2022 (revised) Jaime Jackson

Natural World Publications
P.O. Box 1765
Harrison, AR 72602-1765
www.NaturalWorldPublications.com

ISBN-13: 978-1-7333094-4-8

This natural dental care program is the creation of the author in response to the international epidemic of gum disease and tooth decay. Natural dental care is a non-invasive, drug-free, holistic method of stimulating the person's immune system through diet, exercise, and the application of safe, over the counter herbal and mineral products that have been used by people for countless generations. The author in no way intends or attempts to convey to the reader that he is a licensed dentist, a person trained in the dental sciences, or practices professional dentistry in any capacity. The decision to use this method in conjunction with or without conventional dentistry is the prerogative of the reader. In either case, the author accepts no responsibility for the applications or misapplications of the recommendations in this book.

Contents

Primum non nocere . . . vis medicatrix naturale.
First, do no harm . . . (respect) the healing powers of nature.
Hippocrates (5th Century BC)

Tooth Fairies, Toothaches, and Tooth Truths!

ır earliest childhood memories most of us recall (here in America any-
ıat "giving up" our teeth was a part of life that, ironically, brought re-
gan with losing our "baby teeth." What otherwise might have been a
traumatizing experience of losing such a vital and noticeable part of us, our parents assured us (as a diversion) that a magical "Tooth Fairy" would come during
the night while we slept to reward us with cold hard cash — if only we put that
tooth under the pillow for her to take away. We did, and, sure enough, come
morning we lifted the pillow with great anticipation, and presto, we were loaded
with dough, if only pennies, nickels and dimes! Cash for teeth was the message,
although being the sort of person I am, I did question why the Tooth Fairy
wanted the tooth, where she was going with it, and what she was going to do with
it once she got it there! I recall frowns on my parents' faces. Whatever, I wasn't
the only one among friends talking about the "easy money" and how we could up
the ante on the elusive Tooth Fairy. We soon realized that "nickels and dimes"
didn't add up to much. Further, heated discussions ensued when it was learned
that the Tooth Fairy wasn't exactly equitable. Some of us were getting mere pennies while others were getting the paper stuff. Bartering was even considered as
the dividends were surely greater if we threw in with what might have been a
wealthier or more generous Tooth Fairy. The more desperate among us were tying
strings to the door knobs to yank anything that hinted of loose teeth. But soon we
were told the Tooth Fairy would no longer come, because we were older now and
we needed to keep our teeth. *Sigh.*

Eventually, as we aged, the inevitable tooth aches began. We were then taken
to a person we've never heard of before. *The dentist.* Like the Tooth Fairy, this person was also going to reward us for giving up our aching tooth. But not with nickels and dimes. Those days were over. *By ending our tooth pain.* And, unbeknownst
to us, by mom and dad paying up again — this time actual big bucks. We soon for-
got the experience, sort of, since the money wasn't coming out of our own pockets
and the pain was miraculously gone. But our little gig of forgetfulness was about to
come to an end. And precisely where my story begins . . .

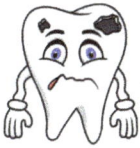

I think most of us have been through the same common "dental care" scenario at some point in our early adult lives. We're told to get our teeth checked regularly by our dentist, brush two or three times daily with American Dental Association's (ADA) approved tooth pastes (ditto with mouthwashes), and eat healthy foods. Not that all of us followed the recommended regimen faithfully, but we tried, and there were many who actually did try very hard. But whether we were slack or diligent, we all seemed to be facing the same outcome in due time, which goes something like this and to this very day . . .

We head off to the dentist for a checkup (without mom because we're now the adults in charge, of the charge cards, that is). We are typically anxious, but hopeful that all will turn out well with our teeth. We're escorted to the examination chair, bibbed, and told the doctor will see us shortly. An x-ray technician arrives, jackets us with a heavy lead apron, crams some kind of device into our mouths to hold them open, and orders us to "hold still" or "hold your breath." We comply. The device is moved to another location in our mouth, several times, in fact, with more x-rays taken. We are then thanked for cooperating so well, told the dentist will arrive in a minute to see us and to share the results. We wait, and wait, and wait.

The dentist finally arrives, offers a curt "Hello, how are you doing," then takes their seat, tilts us back, and tell us to open wide. We do and in goes the tiny mirror and some kind of scraping probe. An assistant has quietly appeared just out of our view to take dictated notes about each tooth in a language we really know nothing about. Finally, the probe is over. We are told there are several "cavities" that need attention right away, others that are "on watch," and that our teeth badly need "cleaning" to remove "dental plaque." *Sigh.* We get the feeling that a very unpleasant but "necessary" pathway into our mouth has been opened.

The 3 Tooth Truths
- Lousy dental hygiene
- Lousy diet
- Lousy genes

Next, we are told to make an appointment with the staff up front so we can deal with the cavities. But we feel compelled to ask the dentist — whom we never really got a close look at, ensconced in the chair as we were facing the ceiling — who is nearly out the door, a few questions about why these cavities happened and if we can prevent them in the future. We are then told of the "3 Tooth Truths": Cavities are caused by inadequate dental hygiene, poor diet, or our genes, or all three. This dental mantra not only sounds reasonable, we are informed to accept it because that's our fate as human beings.

But before leaving, another assistant arrives to clean our teeth. This is done by tediously scraping away "dental plaque" from each tooth, one by one. We're not exactly sure what dental plaque is, except that it is very hard and is stuck somehow to our teeth. But we assume it must be very bad because, we are told, all of it has to be removed and on a regular basis. Afterwards, we are set free. But not completely because we know what has yet to come and we will soon be back in that chair to face the music.

So, not without dread, the appointed day finally arrives. The cavities are repaired with fillings. "It went well," our dentist assures us afterwards, but not another word about the 3 Tooth Truths because it is assumed we have learned that lesson in life and we are to accept it without further doubt or a stream of senseless repetitive questions.

As the years go by, and not too many as is often the case, new cavities arrive and the ones previously filled are now problematic. We are told that they need to be replaced with "crowns"[1] (*facing page*). What the . . . ? Having no other real choice but to walk out, we do as we're told. As more time goes by, we learn that some of the cavities have enlarged and become infected at the root. We are now faced with three choices: remove the entire tooth, undergo a "root canal" to clean out the infection, or do nothing and risk spreading the infection into our gums and only God knows where else. For those who can't afford all the fancy, high tech care, extraction is the cheapest way to go, if not the only way to go if the decay is too advanced. And this will mean anything from a "bridge" to "dentures," or, if money isn't an issue, "dental implants" — with ersatz tooth look-a-likes embedded with screws right in our jaw bones! Nothing to celebrate, and each of these "tooth fatal" outcomes are not — we are warned — without their own set of potentially very serious complications. But the bottom gum-line is made perfectly clear: We're probably going to lose our teeth. Just a few if we're lucky, but for many people, increasingly so, all of them will go. A specialist at extraction is recommended, along with another specialist who will fit us with the ersatz tooth substitutes.

On reflection, if we dare to look back in time and do the math, what is hap-

---

[1]To the extent I'm forced to define dental terminology and explain related research to bring clarity to what is happening to our teeth, my intent is to otherwise keep my language in the common vernacular ("normal talk") in the main text, or, I'll relegate it to a footnote like this if it's important corroborative information to this natural dental care program.

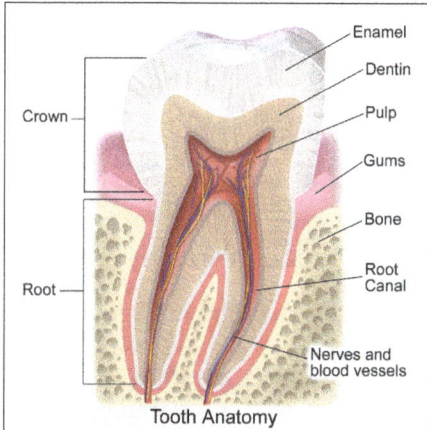
Tooth Anatomy

A *crown* (or cap) is a type of dental restoration which completely caps or covers a tooth. Crowns are recommended when a large cavity threatens the health and structure of a tooth.

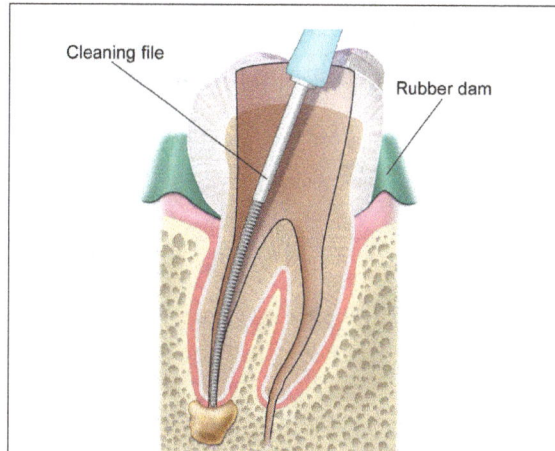
Removing Infected Pulp

Endodontic therapy or root canal therapy is a treatment of the infected pulp of a tooth which theoretically results in the elimination of infection and the protection of the decontaminated tooth from future microbial invasion.

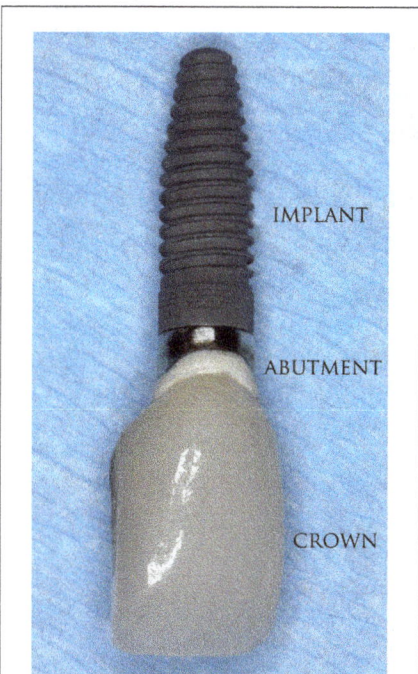
IMPLANT

ABUTMENT

CROWN

A dental implant (also known as an endosseous implant or fixture) is a surgical component that interfaces with the bone of the jaw or skull to support a dental prosthesis such as a crown, bridge, denture, facial prosthesis or to act as an orthodontic anchor. The basis for modern dental implants is a biologic process called *osseointegration*, in which materials such as titanium form an intimate bond to bone.

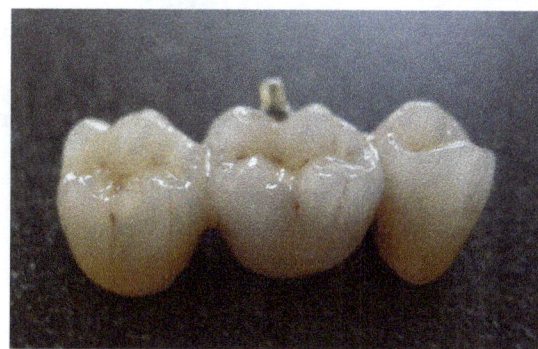

A three unit porcelain fused to metal bridge (PFM). A bridge is a fixed dental restoration (a fixed dental prosthesis) used to replace a missing tooth (or several teeth) by joining an artificial tooth permanently to adjacent teeth or dental implants.

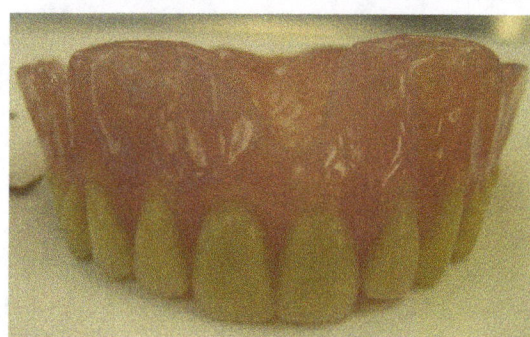

Dentures. Also known as "false teeth," are prosthetic devices constructed to replace missing teeth; they are supported by the surrounding soft and hard tissues of the oral cavity.

pening is that dentistry is really the practice of removing "tooth mass." The industry does this very well, and, clearly, it is very profitable to do so, matching "procedure" to "income." Here's some interesting data to think about:

Worldwide

- Approximately 3.6 billion people have dental caries in their permanent teeth.
- The World Health Organization estimates that nearly all adults have dental caries at some point in time
- In baby teeth it affects about 620 million children or 9% of the population.
- They have become more common in both children and adults in recent years.
- The disease is most common in the developed world due to greater simple sugar consumption and less common in the developing world.

United States

- According to the American Association of Endodontists, more than 15 million root canals are performed every year (41,000 of those performed each day).
- According to the American Academy of Implant Dentistry:
  - ◊ More than 35 million Americans are missing all their teeth in one or both jaws according to prosthodontists.
  - ◊ 15 million people in the U.S. have crown and bridge replacements for missing teeth.
  - ◊ 3 million have implants and that number is growing by 500,000 a year.
  - ◊ The estimated US and European market for dental implants is expected to reach $4.2 billion by 2022.
- According to the National Institute of Dental and Craniofacial Research:
  - ◊ 92% of adults 20 to 64 have had dental caries in their permanent teeth.
  - ◊ 26% of adults 20 to 64 have untreated decay.
  - ◊ Adults 20 to 64 have an average of 3.28 decayed or missing permanent teeth and 13.65 decayed and missing permanent surfaces.

Up until eight years ago (2014), I had, for the most part, accepted this fate and went along with it, if not begrudgingly, then at least hoping for the best. But lingering in the back of my mind was some unresolved doubt about the "3 Tooth Truths." How could nature, I reasoned amid a bit of insecurity, through the powerful forces of natural selection, arrive at such a mouthful of widespread failure in our species?

Well, it really came home to me personally nine years ago when I went to my last and final dental checkup, and from which I walked out without any treatment for anything. I learned that my previous dental work, a healthful diet, a fairly vigorous lifestyle, and diligence with the tooth brush and ADA recommended toothpastes and rinses, did not spare me from a host of new cavities, broken crowns that had failed, one eroding chipped tooth (military related overseas), and one "root" infection that had spread to my gums. Showing me the x-rays of my teeth, which I could not "read" if my life depended on it, but accepting his word that they were "proof," came my brutal marching orders: "You'd better schedule your appointments right away, and it's looking like we're not going to be able to save several of those teeth. And you're looking at several thousands of dollars at the very least, in my office alone." In his office alone? What that meant was I would be in the chair of other specialists for the really involved and costly procedures.

As I left his office, I did the math in my head. I would be completely toothless by the time I reached 75, my age as I'm writing this book. Only, that's not what happened. Two crowns that I had failed with minor infection setting in — so the remnant teeth were lost altogether. Otherwise, I haven't set foot in a dentist's office and the rest of my teeth are in tact. A couple of years ago, I did decided to try (on a foolish whim) a speck of the ADA's recommended mouthwash. I was at a friend's, where on the bathroom counter was a bottle of one of those plaque killing rinses. I thought, what would that feel like in my mouth after all these years of doing what I now call "natural dental care?" Well, only for the purpose of "personal research," I took a small swig. *Holy Moses!* In a flash, my mouth felt like it went into electric shock. It quite literally stunned me. In another flash, I irrigated my mouth with fresh water until my mouth began to calm. The new biological environment within my mouth had revolted and let me know in no uncertain terms it didn't appreciate my little excursion into the past. That aside, the toothaches and infections are now gone, and my teeth, including the ones that lost mass, have actually regained mass in dental plaque (called tartar — more on that later) and are now of a natural color and strength and there are no signs of gingivitis (gum disease) either. In spite of the ravages of my dental past, my teeth and gums feel good.

This book is about my dental healing journey. I hope it inspires and helps

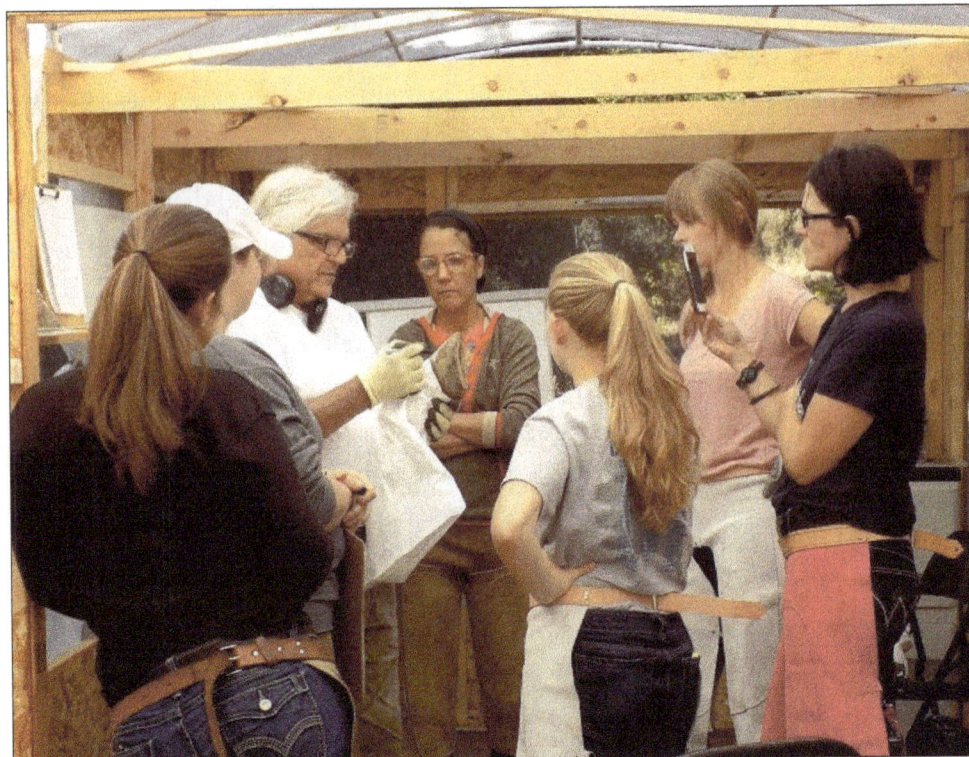

Giving a class on hoof anatomy to students in the ISNHCP Training Program for natural hoof care practitioners at our Field Headquarters in Lompoc, California (USA). Institute for the Study of Natural Horse Care Practices (www.ISNHCP.net).

others to follow suit who are still losing their tooth mass, or who want to prevent serious cavities altogether. We do have choices, and they don't have to be limited to those held over our heads by the dental industry's 3 Tooth Truths.

I'm also including references here and there to my own work in the healing of horses. I've been a professional "horse healer" of sorts for over 40 years, specializing in promoting more healthy lifestyles for people's horses based on my studies of wild, free-roaming horses of the U.S. Great Basin from 1982-1986.[1] Okay, I'm sure you're wondering, why bring that stuff into the picture? The reason is that there are definite parallels in both the science, evidence, positive outcomes, and misconceptions about both of our species in their natural state that are worth sharing. I would go so far as to say this applies to all species in the web of life. But, here, I will address only that which I know definitively about. Those who know me in my professional circles understand I am as much a pragmatist demanding evidence-based outcomes, as an adventurer into the vast, unknown realms of natural healing. So, let us go where no tooth has gone before. Or, at least ours!

[1]Jaime Jackson. *The Natural Horse: Lessons from the Wild.* (1992, 2020).

Chapter One
Our Ancient Ancestors and Their Teeth

At some point in our own species's ancient (paleolithic) history, each one of our earliest relatives, *Homo sapiens sapiens*, probably walked out of a cave to greet the day. One thing, for sure, none of them were heading to the dentist's office. Not because there weren't any, but because there was no need for them. Scientists are now unearthing new evidence in the archeological record that our ancient ancestors had perfectly healthy teeth. Tooth decay simply wasn't a problem like it is today. One might go so far as to say that dentists would have been as irrelevant as a salesman trying to pitch ice to an igloo dweller at the dawn of humanity. I've pondered this new data and have asked myself, wouldn't this be enough to give the dental industry pause long enough to question the 3 Tooth

Truths? Well, if not the dentist too busy drilling away, then how about the researchers with their science that underlies the professional dental industry? Whatever it is that happened to our ancient ancestors with their perfect teeth that has led to today's massive epidemic of tooth decay is incredibly important information. Because genetically they are still a part of us. *They are not irrelevant.* The more we can learn from them, the more we can learn about ourselves — including our rotting teeth. I ask, "What went wrong?"

One thing, for sure, they weren't heading to any dentist's office!

We can now trace our genetic roots directly to them. Think about it. It is remarkable that until very recently, no human today could trace their family lineages back in time to the mysterious and shadowy origins of our species. But modern genome science is connecting our heritage dots to them in ways we never could have thought possible less than a few decades ago. They were humans, our species, our ancestral families from which sprung all our relatives and friends in our lives today. During their earliest time on earth they had no fire, no fancy technology like we enjoy today. They had families, loved ones no doubt they cared about, or they would have delivered themselves into extinction. They understood that survival meant innovation, and so they changed in the name of progress as they would have understood the concept to mean in their time. But modern science is now opening new doors into their ancient lives — before they became "civilized" with tooth decay. And there is good news! Science has discovered that our ancient ancestors passed down to us an adaptation of great teeth. Yes — "great teeth." This is nothing to belittle, ignore, or deny. And they have given us vital clues as to how it all worked for them, and more clues are coming as new, unprecedented science unrelentingly pursues our most remote past. But it is not all good news, unfortunately, as humanity marched closer and closer towards the present.

For example, in one scientific forum on tooth decay, we learn, "There is also evidence of caries increase in North American Indians after contact with colonizing Europeans. Before colonization, North American Indians subsisted on hunter-gatherer diets, but afterwards there was a greater reliance on maize agriculture, which made these groups more susceptible to caries."[1] The problem with this confusing statement is that it suggests that the maize (corn) culture was "bad" for teeth, and somehow arose after the arrival of the "white man" — replacing the hunter/gatherer lifestyle. But that's not what happened. Corn culture was well established in North America as early as 4,000 years ago, and widely so across the United States 500 years before Columbus arrived in the New World. In fact, Native Americans introduced early colonists to corn, tomatoes, potatoes, squash and many other indigenous edibles across the "New World." Their hunter/gatherer lifestyle ended for the same reason that their maize culture also ended. It is equally well-known that the diets of virtually all Indian nations, in fact, of "primitive" peoples worldwide, were altered with devastating outcomes as their ancient connection to the natural world was supplanted by the pernicious effects of militarily enforced colonization.

But, here precisely, my point is that while dental scientists may point to the fact that the industrial forces of colonization screwed up native people's diets, it is almost reckless to ignore that many, probably most, engaged in some sort of agriculture prior to contact. We can't say they were "hunter-gatherers" while pointing to corn culture as the agent of dental caries after contact, when neither statement is true. And what I have found also is that this strain of negligent science permeates and infects much of our knowledge about the actual diets and dietary habits

[1]Epidemiology of Dental Disease, hosted on the University of Illinois at Chicago website. January 9, 2007.

of our ancient antecedents and the true health of their teeth. Not good!

Nor should we really leave it at that, because this negligence factor actually infects the modern dental industry's assumptions about natural diets and their relationships to healthy teeth. It is embedded in the false narrative of the "3 Tooth Truths." This is not to say there isn't valuable data amid the deplorable research that is available to us, thank God. It's just that you have to be discerning of *bias*. Speaking of bias, I would go so far as to say that the entire body of scientific and historical literature on oral disease and oral hygiene among ancient peoples is so polluted, that, with very few exceptions (which I will bring up shortly) all of it is suspect. In fact, both past and present research often oozes arrogant, racialist bias towards "primitive" peoples and their diets. Some research is just plain bad, not well thought out or carried out – unless, we must assume, bias was their intent! I am astonished by the indifference to nature that has come to jade so much of modern scientific research. Who is conjuring up all of this, and, how in heaven's name does it get funded in the first place? What is particularly distressing is when bad science is adapted by dental "experts" to explain the pathogenesis (origins) of oral disease based on diet (Tooth Truth #2). They're just wrong (but not all of them!). A lot of this misinformation — like the Native American corn culture myth — borders on incestuous science, with one scientific journal passing it along "in house" to the next, university to university, and citing it as academic research which they can then all embrace. But it doesn't end there.

Modern science is also, I have to point out, biased towards its profitable industrial outcomes, which, in turn, are beholden to their investors. No profit, no research. To me this is a critical watershed between limited, even corrupted, science and progressive science based on critical thinking and the genuine yearning to explore, discover, and understand, for example, the laws of nature as they apply to nutrition. One line of investigation pursues solving systemic problems plaguing our vitality due to unnatural diets, but solely on the basis financial reward. The other simply seeks to know why a dietary practice by "primitive" peoples has few or no inherent problems, while investors are running in the opposite direction!

This scientific divide inevitably clashes in the realm of utility, and nowhere is this division more amplified than with modern dentistry versus what I've come to know and practice as "natural dental care." Dentists whom I've attempted to discourse the relevance of ancient genetics, natural healing mechanisms, or even stem cell technology, simply cannot go there except to mock any such discussion derisively or be dismissive on the grounds of "no credible scientific evidence." They are bound, probe and drill — like an auto mechanic with a tire iron — to do what they do and that is the end of the discussion until science (based on financial rewards!) comes down from above to tell them what this next profitable direction is they're all going to pursue. I find this myopic rigidity to be alarming, especially when so much tooth mass is at stake. That's our tooth mass, yours and mine, and my purpose here is to protect it from their drills. I'll return to this dentistry impasse in Chapter 5, but let's lay a little more groundwork first so we can understand why it is so, at least from my perspective.

Most people (including dentists) would logically assume that teeth in the natural world of our earliest ancestors simply had to be problematic based on what we have seemingly inherited from them in our own troubled mouths (Tooth Truth #3). The data cited in my introduction also supports this perception. In defense of the dentist, I can't imagine that a serious alternative Paleolithic view is brought up at any dental school based on the abundance of flawed research and the entrenched mantra of the 3 Tooth Truths. And so modern dentistry, logically, drills its way along without reason for doubt. But now there is reason for doubt due to new strain of scientists who have looked deeper into our species's ancient genomic history. For those manning the archeological digs, it is a credit to their critical thinking, intuitions, and good science - that they are trying to understand what they are actually seeing and honestly reporting the facts - including the facts about our own specie's remarkable dental "truths." Even if their work is at loggerheads with an entrenched dental industry that would have us believe otherwise.

It is our good fortune that these scientists have begun to trace our specie's teeth across a timeline spanning over 350,000 years, from the dawn of *H. sapiens sapiens* to the present "modern" human — that's "us!" Those who have looked hard and honestly so, in my opinion, have arrived at an informed opinion based on credible evidence that the earliest humans of the Paleolithic Era ("Stone Age Hunter-Gatherers") had healthy teeth. But as time marched forward, somewhere around 10,000 years ago in the Neolithic Era ("Farming and Domestication of Animals"), more and more humans began to settle down in agricultural cultures.

And as they did, leaving behind their hunter-gatherer lifestyles, so researchers note, they began to have dental issues – very serious tooth decay and gum disease. Why?

"Hunter-gatherers had really good teeth," asserts Alan Cooper, director of the Australian Center for Ancient DNA. "Apparently, our ancient ancestors ate lots of wild animal flesh as well as a broad range of plant life, but had no agriculture, [but] as soon as you get to farming populations, you see this massive change. Huge amounts of gum disease. And cavities start cropping up."[1] Cooper and his research team studied calcified plaque on the teeth of 34 prehistoric human skeletons. What they found was that as ancient diets shifted from meats, vegetables and nuts to processed carbohydrates and sugars, so did the composition of bacteria in their mouths. Bacteria, as it turns out, that create big problems for teeth. In fact, as much is no mystery and is well-documented today, and even quoted astutely if need be, by our dentists, who point out that these bacteria metabolize sugars into free acids that erode teeth and cause cavities.

There is an interesting parallel between the advent of tooth problems with Neolithic humans and the contemporaneous extinction of wild horses in the Western Hemisphere. Sugar-loving bacteria appear to be at the bottom of both scourges. I have long held in my own books that late-Pleistocene populations of wild horses could not adapt to Fructan (sugar)-rich grasses left in the wake of melting glaciers across North America. It is worth noting here that these wild horses (*Equus ferus ferus*) are genetically identical to today's modern horse (*Equus ferus caballus*), who are also sugar intolerant. In fact, NHC science classifies the horse as

[1]I was recently surprised to learn that Cooper happens to be a good friend of one of my clients on whose property in Canada, some of his research on the DNA of recently discovered ancient horses has taken place. Cooper's ground-breaking research may be accessed from the Australian Centre for Ancient DNA, School of Biological Sciences, University of Adelaide, South Australia 5005 Australia, and online at http://www.adelaide.edu.au/directory/alan.cooper. Pertinent to my own investigations and the thesis of *Guard Your Teeth*:

Weyrich, L., Duchene, S., Soubrier, J., Arriola, L., Llamas, B., Breen, J. ... Soltysiak, A. (2017). Neanderthal behaviour, diet, and disease inferred from ancient DNA in dental calculus. *Nature*, 544, 7650, 357-361.

Weyrich, L., Dobney, K. & Cooper, A. (2015). Ancient DNA analysis of dental calculus. *Journal of Human Evolution*, 79, 119-124

Adler, C., Dobney, K., Weyrich, L., Kaidonis, J., Walker, A., Haak, W. Cooper, A. (2013). Sequencing ancient calcified dental plaque shows changes in oral microbiota with dietary shifts of the Neolithic and Industrial revolutions. *Nature Genetics*, 45, 4, 450-455.

Adler, C., Haak, W., Donlon, D., Cooper, A. & Dersarkissian, C. (2011). Survival and recovery of DNA from ancient teeth and bones. *Journal of Archaeological Science*, 38, 5, 956-964.

Adler, C., Haak, W. & Cooper, A. (2009). Quantification of DNA in ancient human teeth. *Homo-Journal of Comparative Human Biology*, 60, 3, 262-262.

Cooper, A., Rambaut, A., Macaulay, V., Willerslev, E., Hansen, A. & Stringer, C. (2001). Human origins and ancient human DNA. *Science*, 292, 5522, 1655-1656.

Wild horses and other megafauna of the late Pleistocene all faced extinction in the Western Hemisphere. Is there a scientific explanation that also explains tooth decay and other diseases in us?

an insulin-resistant (IR) species, much like the human diabetic who cannot metabolize sugar. What research has revealed today – and which, in fact, parallels the plight of our Neolithic ancestors — is that conventional grass pastures rich in sugar and carbohydrate enriched horse feeds cause shifts in the digestive biology of the horse that favor aggressive, potentially harmful strains of bacteria, notably, *Streptococcus bovis* and *Streptococcus equinus*.[1] These gram-positive bacterial strains precipitate an acidic shift in the pH of the horse's intestines, which erodes the mucosal lining allowing contamination to uptake into the vascular system. Normal circulation then transports these harmful strains of bacteria and their toxic waste matter throughout the body.[2] In my book, *Laminitis–An Equine Plague of Unconscionable Proportions*,[3] I explain how this pathophysiology (functional progression of a disease) devastates the horse's foot and other organs of the body. Cooper's archeological findings and interpretation of his data independently parallel my own. Not surprisingly, our recommendations for doing something about tooth decay and hoof disease are also in parallel concert. I'll knit the two together in Chapter 3. We still have more foundational material to cover before we get to that discussion.

---

[1]*Equine Laminitis*. C. Pollitt, BVSc, PhD; M. Kyaw-Tanner, BSc, PhD; Kathryn R. French, BSc, PhD; Andrew W. van Eps, BVSc; Joan K. Hendrikz, BSc; Mousa Daradka, DVM, PhD. Australian Equine Laminitis Research Unit, School of Veterinary Science, Faculty of Natural Resources, Agriculture and Veterinary Science, (Pollitt, Kyaw-Tanner, French, van Eps) and School of Medicine, Faculty of Health Sciences (Hendrikz), The University of Queensland, Brisbane, QLD 4072, Australia; Department of Veterinary Clinical Studies, Faculty of Veterinary Medicine, University of Science & Technology.

[2]*S. bovis* has also been found in humans and is implicated in colon cancer. I will discuss this further later in the book.

[3]*Laminitis: An Equine Plague of Unconscionable Proportions – Healing Your Horse Using Natural Care Principles and Practices*. (2016) J. Jackson. NHC Press, p. 19.

Chapter Two
Industrialization of Our Food Chain and the Rise of Dental Caries

My opinion is that good research has established more than a justifiable argument that maybe we shouldn't be pointing to genetics — pathological mutations aside (i.e., Tooth Truth #3). Following this line of deductive reasoning, we are left with diet and oral hygiene. Here, conventional data underlying both is so convoluted and murky with unsubstantiated claims and corporate PR, it would be almost impossible to decipher the remaining two "Tooth Truths" were it not for an abundance of good evidence coming from the dental industry itself!

There's not a dentist hanging over our mouths with drill in hand who hasn't shown us dreadful photos of patients' decayed and missing teeth (e.g., see page 76), and who don't own tooth brushes, don't brush their teeth nearly enough (if at all), drink soda nearly 24/7, smoke, don't exercise, and eat fast food until they're on statins to control their high blood pressure and "bad" cholesterol levels. It's a scary sight, to say the least! And good propaganda to keep us in line at the dental office, where we're still ensconced in that examination chair.

While I'm no longer personally impressed or intimidated by this terroristic dental tactic, like the whirring drill itself, there is something we can glean from what they are saying that does make sense. And both Cooper and I would surely agree! As our Neolithic ancestors took stock of the land and began to find ways to be more efficient and productive, they also found a way to make it increasingly profitable. I suppose human nature being what it is, this comes as

Ascent or descent of healthy food and great teeth?

no surprise. From the earliest farming in Mesopotamia thousands of years ago, to the advent of feudalism in Europe, to the triangular slave trade of the 19th century, and on to the 20th Century Space Age Industrial Revolution that eventually landed us on the moon, today the burgeoning corporate amassing of wealth and power has determined the very composition of our food chain, our diets, our dental care, and the bacteria inhabiting our mouths! It's simply undeniable. And down through the centuries, people haven't always been particularly happy about this type of "evolution" either. In my book *Laminitis: An Equine Plague*,[1] I wrote:

> Late 19th century Victorian concerns about disease and human indigestion are also well-documented.[2] These concerns arose almost as an obsession, if not outright hysteria, because of the perceived impact and dangers of an emerging industrialized food chain. Scientists, physicians, patients, and the general populace alike all spoke out fearfully against what was happening — people were no longer eating "naturally" and "healthfully." One could argue that the modern "natural foods" movement began then — and with good reason! In fact, I personally share those same concerns today and I'm glad to see the emergence of the modern "organics food" movement. Because of organic farmers, ranchers, and processors, we have at least a little edge over industrialized foods that are chemically cooked, bound, and found everywhere.

But my point here is that today's international "natural foods" movement speaks directly to the problems people experience with the industrialized food chain. Dentists will tell you modern life can make it difficult for busy people to eat healthful foods. And the poorer you are on the economic ladder, matters go far beyond busy schedules to the issues of access and affordability. There is little debate that a less than natural diet of wholesome foods will compromise our teeth and overall vitality, and this extends to the dietary transition between Paleolithic and Neolithic peoples.[3] On the other hand, what today actually constitutes a "natural, healthful diet" is itself nothing less than a burning hot topic among warring factions in the natural foods movement.

As an example, within the broad spectrum of "natural foodists,"[4] vicious battles take place on the Internet concerning the content and viability of purported

---

[1] *Laminitis: An Equine Plague*, p. 93.

[2] *How the Mid-Victorians Worked, Ate and Died.* Paul Clayton and Judith Rowbotham. Int J Environ Res Public Health. 2009 Mar; 6(3): 1235-1253. Published online 2009 Mar 20. doi: 10.3390/ijerph6031235. PMCID: PMC2672390

[3] Molleson TI, Jones K, Jones S. 1993; Dietary changes and the effects of food preparation on microwear patterns in the Late Neolithic of Abu Hureyra, northern Syria. *J. Hum. Evol.* **24:** 455-468.

[4] *Foodist* – A person who discriminates against other people because of the food they eat!

"Paleo" diets. Taking their cue from our Paleolithic ancestors, this means lots of meat, fat, nuts, berries, and raw this and raw that. But this is nothing less than a slap in the face of the gentle, earth friendly vegetarians and vegans, and a tantamount declaration of war on vegetarianism and veganism.[1] And while you would think one battle line would be drawn, each side is cleaved itself into numerous factions, who can't agree on anything either. Amid this food debacle and near gang warfare, industrialists watch closely to see who and what they can exploit on both sides in the name of profits. For the most part, you will find the industrialist's versions in supermarkets, the naturalists in the health food stores, and everyone in sight probing the Internet. This is as far as I will take the discussion on either side, because the task of sifting through their positions and counter positions is no less daunting than sifting through the archeological record dug up by scientists studying our ancient ancestors' skulls and teeth.

But what I will say is that our DNA today is no different than that of our earliest Paleolithic ancestor, *H. sapiens sapiens*, whom scientists have shown had few or little of the dental problems that plague us today. What this means is that each of us will have to pick axe our way to our own truths in view of that evidence. And our own teeth will reveal the truth of the matter. Hopefully, not at the end of the dentist's drill. But not to end on such a dour note, there is an important piece to the puzzle as we configure our diets: The record has shown also that nature has selected us as *omnivores*. Indeed, our species did very well dietarily with what nature provided for our primordial adaptation. And surely without an iota of "political correctness" during our first dinner table grunts. More on diet later.

---

[1]Being an omnivore, I've been lectured a time or two hundred by born-again vegans and vegetarians concerning the inhumanity of my dietary habits. So confronted, I've naturally wanted to know what these eating disciplines are, what differentiates them, and why they've got it in for me. At the simplest possible level of interpretation, vegans forsake eating anything derived from animal products, although they clash vehemently among themselves over whether or not humans should have any relationship with animals, including owning pets. It appears that veganism arose out of an ancient division in vegetarianism, although factions on both sides may disagree with me. So far as I can tell, and I'm daring to walk the line, there are two kinds of vegetarians — the vegans, described above who eat no animal food products whatsoever; and, those excommunicated vegans who have fallen from grace as they do not object to eating eggs, dairy products, or fish.* They also love their dogs, cats, and horses. All factions and all sides are particularly aghast by the red meat eating habits of Paleos and omnivores like myself. Zoos, you can imagine, are practically the end of the world for vegans, the mere idea of which gives them great anxiety, and which fuels the donor ranks of PETA and other veggie radicals. But there's hypocrisy in both vegan and vegetarian ranks — I've caught red-handed more than one eating hot dogs and other tabooed fleshy delights in what were clearly clandestine culinary excursions into my world. Whatever the case with one's eating habits, tooth decay has shown no safe quarter.
[*For an early historical perspective: "International Health Exhibition", *The Medical Times and Gazette*, 24 May 1884, 712.]

Chapter Three
How and Why Tooth Decay Occurs

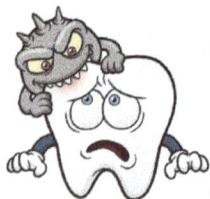

We are told that our teeth decay from the outside inward towards the sensitive tissue within. Diagrammatically and photographically, we've all seen the imageries posted in the dentist's office walls, although we don't wish to think about any of it, and of that I'm completely sympathetic. I don't either! Further, we are told that as the decay approaches the nerves ensconced within the tooth's innermost root cavity, we will begin to feel pain. And we do! Nature's warning that something going on in there isn't right. In other instances, the decayed and weakened tooth may just crack and give way altogether. Not good news at all!

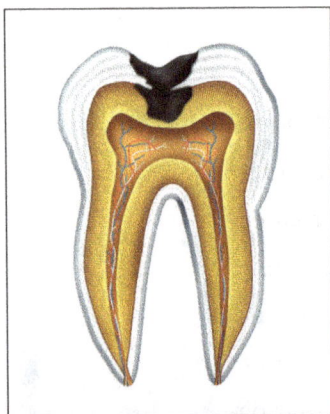

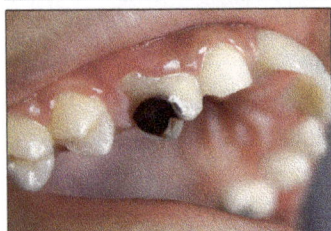

Historically, beginning with our Neolithic ancestors — so the archeological record has shown — few of us have not felt the proverbial "tooth ache." The anthropologist Robert Heizer described how California Indians used bow drills, similar to their fire making tools, to drill into their teeth to root out the decayed matter in hopes of alleviating the pain. The same "dental drill procedure" was apparently used in Pakistan 9,000 years ago on decayed molars. Paving the way for modern dentistry, these Neolithic "dentists" also learned to pack the cavities, and beeswax was used as a dental filling 6,500 years ago in Slovenia. Others sought to deal with their dental pain more definitively. I can't recall the name of the famous Plains Indian warrior chief, who rode into battle declaring, "It's a good day to die," as his toothache was killing him with pain and he clearly meant death in war at the very least meant lasting relief! Stories like these fill the anthropological record of Neolithic "primitive" societies dealing with tooth decay and associated pain.

Former Paleo diets were either altered by the agricultural revolution or, in more recent centuries, corrupted with great devastation by European colonization in the wake of the Age of Discovery. But some did manage, temporarily anyway, to escape this clash with civilization, and astute anthropologists have been there to record what they saw from diet to excellent tooth health. Indeed, there are other records — ignored by the burgeoning dental industry hungering for your teeth — that are revealing of indigenous peoples of North America, Africa, South America,

Why are these People so Healthy?

Native people eating traditional foods had physical excellence, splendid facial and dental arch forms, and no cavities.

Images of healthy teeth according to early 20th century research by Weston Price.[1]

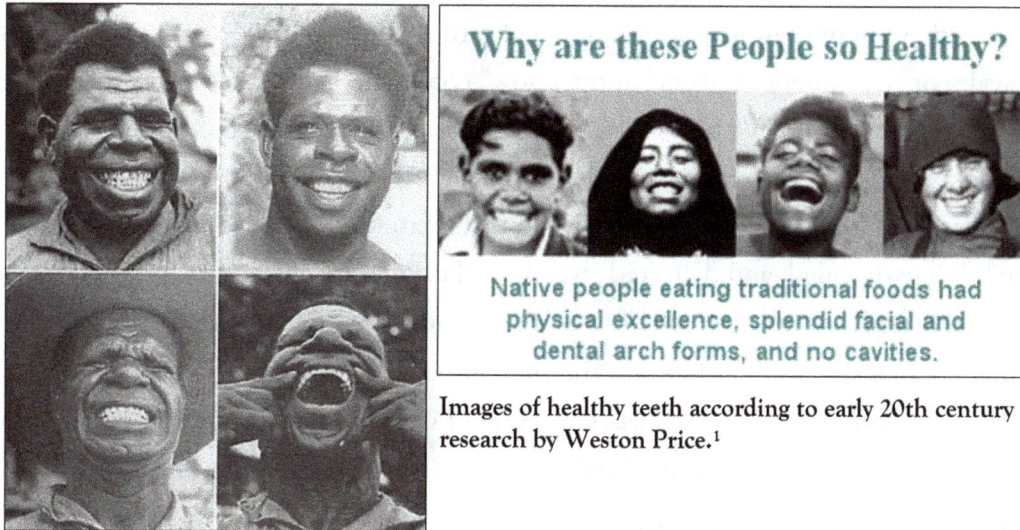

the Arctic north, Mongolia, and Australia, all having flawless teeth at the time of contact with European explorers — and as late as the early 20th century (*above*). What all of them have in common is having had minimal or no contact with the ever expanding industrialized food chain and institutionalized dentistry. But those peoples' lifestyles are now gone. Their ancestors' diets have been completely eradicated and supplanted by what we all eat today, and, for those who can afford it, trips to the dentist. Meaning, they're enduring flawed teeth now with the rest of us.

So, what has happened to our teeth exactly? What are we eating that's so terrible that it's rotting our teeth here in the modern Information Age? How does decay happen? And is there anything we can do about it? There is, but it will take some personal initiative. And the dental industry, as it now commands things, just isn't going to help us. Mainly because they don't really know what to do, con-

---

[1]*Nutrition and Physical Degeneration: A Comparison of Primitive and Modern Diets and Their Effects.* Weston A. Price, MS., D.D.S., F.A.G.D. Member Research Commission, American Dental Association; Member American Association of Physical Anthropologists; Author, *Dental Infections, Oral and Systemic.* Foreword by Earnest Albert Hooton, Professor of Anthropology, Harvard University,; With 134 figures. Paul B. Hoeber, Inc. Medical Book Department of Harper & Brothers, New York London. Copyright, 1939, by Paul B. Hoeber, Inc. [Price, who died in 1948, was extremely controversial in his day and right into the present, and the dental industry then and today dismisses most of his research and related practitioning as quackery. While his contention that foods stripped of their vital minerals and vitamins explained dental disease has some obvious merit, it's important to understand that dental science had not yet implicated specific acid-metabolizing bacteria and their causal relationship to dental caries, even though *Streptococcus Mutans* had been identified by 1928. Not surprising, Price is as popular as ever today among Paleo foodists who credit him with pointing to the healthy, decay-free teeth of native peoples spared the ravages of our industrialized diets, with which I concur — although not based on Price's very flawed and blatantly biased research. He's also the inspiration for modern "holistic dentistry," which I want no part of. Both Price and holistic dentists of today are as invasive of the oral cavity as any mainstream dentist. If Price pioneered anything revolutionary that I would define as "natural dental care," I've yet to see it. Further, I believe his ideas on nutrition were as faulty as his dental practice. I'd of stayed out of his chair knowing what I know now. — J. Jackson]

firmed as they are in their mythical "3 Tooth Truths" mindset. So, it's up to us. There's no cavalry of sympathetic dentists "in the know" coming to save us. The only thing coming at us are their drills. We have to begin with a clear understanding of how and why tooth decay occurs to keep their drills out of our mouths. We must rally around our battle cry, "Guard Your Teeth!" In this effort, we must be careful not to buy into the dental industry's "signed, sealed, and delivered" word salad on the science of dental caries (tooth decay), and, for that matter, much of their bogus and biased science. Going into this discussion, it's best that we bear in mind the very premise of this book: *How could nature, through the powerful forces of natural selection, arrive at such a mouthful of widespread failure in our species?* Well, just maybe nature didn't.

The "sugar is bad for you" paradox

I grew up thinking, like most everyone else on this planet today, that sugar and soft drinks (and other sweetened junk food) rotted or dissolved our teeth. I even recall an experiment where someone poured Coca-Cola over a metal nail in a glass, and not too long afterwards the nail was completely dissolved. Pretty convincing! But, then, I thought, wait a minute, I've smothered my teeth in lots of sugary items and drenched them in soda pop more times than I can recall. And I didn't see the teeth dissolving away like that metal nail. Somehow, things just weren't adding up according to what I was told and seeing with my own eyes. Well, the fact is, you can paint your teeth with sugar all day and all night and the sugar isn't going to dissolve anything. Try it, and you'll see. And by the way, if we just swish it away with warm water – which will dissolve any sugar — isn't that getting rid of it as a cause? True. But, if sugar does cause tooth decay, then how does it do it?

Swishing dissolved sugar around in your mouth in and of itself, doesn't erode your teeth because it's not an acid. The stuff in that can of Coke that dissolved the nail provided the acid (Coke's pH is reported to be 2.6 to 2.7, mainly due to H_3PO_4, phosphoric acid). But back to the sugar: It's what our body metabolizes the sugar into that destroys our teeth (and not just our teeth as this book will show). And that's how you're going to get that cavity (aided by sweetened colas). And, as your dentist will weigh in to add, "Once the tooth is melted away, you're not going to get it back because the body can't grow a new tooth. So get back in my chair." That's a pretty discouraging scenario, to say the least.

The actual *pathophysiology* of tooth decay (how tooth disease progresses) is a separate discussion from its *pathogenesis* (what causes tooth decay). But both need to be understood on some level if we're going to stay out of the chair. Conventional dental science interprets and treats tooth decay as a disease of the oral cavity, much the same way that equine laminitis is classified as a disease of the horse's foot — and it's just as nasty as tooth decay. For years, I accepted this categorization of both human oral and equine hoof disease. But that changed as I begin to investigate causality, first with horses, later with our teeth. Today, I interpret and treat both "hoof and mouth" diseases – and other metabolic diseases – as symptoms of what I call "Whole Human (or Horse) Inflammatory Disease," (WHID). You could also call it "Whole Body Inflammatory Disease." Take your (tooth or hoof) pick! Natural dental care is principally concerned with targeting and arresting causality (WHID) to effect a "healing," whereas symptoms are treated palliatively to mitigate inflammation, meaning temporary "relief" from pain while our bodies are busy healing. Conventional dental care, in contrast, treats symptoms to bring relief, but not causality which is relegated to "this, that, all, or none" rhetoric of their 3 Tooth Truth mythology. Similarly, conventional veterinary care also treats inflammatory symptoms of the horse's foot (called laminitis), but not causality (WHID).

This distinction is important because "treating symptoms" lies at the forefront of both human medical/dental and animal veterinary care today, and, arguably, both default to the willful neglect of treating causality. Effectively, this dental neglect borders on calculated malpractice from my perspective, as it sustains chronic oral disease, evidenced by the mega-quanta of tooth mass ground away into oblivion every year. Certainly this is the case with the modern epidemic of human oral disease and equine laminitis (let us not forget our savaged four-legged friends also taking a beating), both of which are clearly out of control like a raging inferno that has breeched every fire line in its path. Toothaches and inflamed hooves aren't fun, and both, as I will show later, can result in death.

To my way of thinking (and as an equine healing practitioner), we can keep on treating hoof and oral diseases as symptoms, or we can tackle them at their root cause and really bring them under control. Unfortunately, treating diseases as symptoms is profitable, whereas rooting out their causes isn't. *The money, clearly, is in pathology — not health and vitality.* And when matters get very painful and not particularly pretty to look at (rotted or broken teeth), people will go for palliation and put prevention on the backburner. But, for those of us concerned with saving our teeth instead of profits, let's focus on doing what nature intended us to do all along, rather than stay stuck to the dental regimes of profiteering at the expense of our teeth and unpleasant lives in the chair.

We and our "microbiota"

Foundational to natural dental care is the understanding that nature has built our bodies with vast and complex communities of microbiota (microscopic organisms) that are intended to live together symbiotically if the whole lot of them (and us as hosts!) are to survive.[1] An important part of that symbiosis is how everyone gets fed. Feed the wrong things, and the microbiota invariably turn upon themselves in what can only be described as vicious, self-serving warfare to survive at all costs. Warfare that we pay for with menaced bodies and teeth! Back to Cooper:

> What you've really created is an ecosystem which is very low in diversity and full of opportunistic pathogens that have jumped in to utilize the resources which are now free . . . and that's a problem, because the dominance of harmful bacteria means that our mouths are basically in a constant state of disease. You're walking around with a permanent immune response, which is not a good thing . . . it causes problems all over the place.[2]

In this interpretation, we can only see ourselves as our own worst enemies since we're the ones doing the feeding. If we're talking about microbes in our mouths that are getting out of hand, the temptation is to wage our own war in there *on them*. Exactly what the dental industry want us to do! And I can sense you're getting ready to make an appointment right now to get back in the chair. Well, don't!

As it turns out, nature has seen fit to colonize our mouths — and everywhere else in our bodies — with a diverse spectrum of bacteria (and other micro-critters we need not concern ourselves with — at least for now!) that are adapted to live there together in harmony. Only, it's a very tenuous harmony and our industrialized food supply has given some of them, truly aggressive nasty ones – *Streptococcus mutans* and *Streptococcus sobrinus* among the most virulent – the advantage over the others when it comes to many refined foods we consume. What this means exactly for our teeth, I'll go into shortly. For clinical purposes of targeted eradication they are often considered together as a group, called the mutans streptococci. On the other hand, *Streptococcus sanguinis* is another species of bacteria of the *Streptococcus* family that also inhabits our oral cavity, but acts to modify that environment in such a way as to help other mechanisms operative in our mouths to moderate the colonies of mutans streptococci. This fact is significant and will play an important role in restoring our teeth to health, discussed later in this book. But before we get into that, I need to put the brakes on where you're probably going in your head, as

---

[1]*Symbiosis* — the living together in more or less intimate association or close union of two dissimilar organisms.

[2]Adler, C., Dobney, K., Weyrich, L., Kaidonis, J., Walker, A., Haak, W., Cooper, A. (2013). "Sequencing ancient calcified dental plaque shows changes in oral microbiota with dietary shifts of the Neolithic and Industrial revolutions." *Nature Genetics*, 45, 4, 450-455.

I've been down that road too.

To be accurate and clear, I'm not suggesting here that we're talking about "good" and "bad" bacteria. What I mean is that we're talking about colonies of bacteria whose ecology is so disrupted by our industrialized food chain and dentistry (which includes an allied dental pharmacopoeia), that the outcomes of that disruption lead us to believe that the mutan streptococci are the quintessential "bad guys" — marauding gangs of tooth decayers — while *S. sanguinis* are the "good guys" coming to save us. No one's coming to save us. It's down to us and that ecology we've got within us to balance if we are to save the day. In other words, all those bacteria are meant to be there, just in a "balanced way." The Sioux Indian healer, Little Crow expressed this sentiment better than anyone I know in his book, *Sacred Hill Within – A Lakota World View*, aimed at helping American Indians struggling with alcohol addiction: "There are no outs, no scapegoats, no fall guys, no more saviors for us. Now, it is down to just us, you and I. We did it and we have to clean it up, stop it, change it, or let it go on as it is. No one, and I mean no one, is going to come down from anywhere and save our asses."

So, these discombobulated bacteria, neither good nor bad, are who we've got to bring together in a "grand harmonic convergence." To save our teeth! But before we can bring them together so everyone's happy, or at least nutritionally satisfied, we have to know what they want and need to eat, and also what their natural "roles" are in our mouths in the first place. If we can't answer those questions, believe me, our ship is sunk. And with half our teeth already degraded or gone (I hope not!), and the dentist on the way to get the rest of them, we can't exactly look into our own disastrous mouths to see who should be doing what. It's chaos in there! But, we can at least see what they're doing that's creating the chaos — with the dentist's help and a lousy diet, of course. And work our way out from there to a better understanding. I see no other way to go about this.

§

Collectively, these colonies of bacteria form a sticky, clear, living layer called *biofilm*, and it's in and about everywhere on our teeth and gums. As it forms, you can actually feel it with your tongue — I would describe it as "slime-like," others say "fur-like." This bacterial growth is perfectly natural, and nature intends it to be there. And on this point precisely begins our great divide from conventional dentistry. Biofilm is what the dentist and the entire dental support industry tells us is

Association	Layer of saliva provides surface substances for bacterial attachment.
↓	
Adhesion	Within hours, bacteria begin to attach themselves.
↓	
Proliferation	Bacteria multiply within mouth.
↓	
Microcolonies	Diverse microcolonies of bacteria secrete protective layers ("slime").
↓	
Biofilm formation	Microcolonies assert their unique advantages for plaque stability.
↓	
Mature colonies	Biofilm establishes a healthy, balanced proliferating community.

Simplification of natural plaque formation.

"plaque." In fact, dental science is very familiar with biofilm and call it by various names, including microbial plaque, oral biofilm, dental biofilm, dental plaque biofilm and bacterial plaque biofilm.[1] By whatever name you call it, it's the stuff that the dentist — we recall back in the chair in the Introduction — tells their assistant to scrape off. Natural dental care takes the opposite position: Leave it alone. I know, I know, you're really getting nervous now. Read on!

Not surprisingly, plaque formation is incredibly complicated and a true wonder of the natural world (*see simplified flow chart above*). Nature has provided an amazing scaffold of biological mechanisms to prepare and stabilize the bacterial colonies that, through natural selection, are adapted to take residence in our mouths. *This adaptation took place over 300,000 years ago at the dawn of our species.*[2] Our saliva plays a key foundational role in providing a stable foundation for bacterial adherence to our teeth and regulation of bacterial colonization in the many specialized niches they occupy symbiotically in and around our dental arcade. Much science has targeted plaque formation and an entire book could be written on the subject alone. A part of me wants to do that here because it is of great interest to me. But I won't because it isn't necessary and such a discussion is potentially so tedious as to destroy all interest from my targeted audience — that's you

[1]Darby M L, Walsh M M. *Dental Hygiene Theory and Practice.* 2010
[2]Callaway, Ewan (7 June 2017). "Oldest Homo sapiens fossil claim rewrites our species' history". *Nature.* [The authors state 315,000 years ago, but as more reliable DNA sources arrive from new discoveries in the field, that date may be driven back even further, since it is believed by other scientists that "functional modern human DNA" diverged from Neanderthal hominids over 500,000 years ago, and possibly much later than that — J. Jackson citing: Green, R. E.; Krause, J; Ptak, S. E.; Briggs, A. W.; Ronan, M. T.; Simons, J. F.; et al. (2006). *Analysis of one million base pairs of Neanderthal DNA. Nature.* pp. 16, 330-336

reading this! So, what follows is a "brief" of what's going on and its significance for our natural dental care program.

§

Stabilized biofilm is drenched continuously in various minerals, such as calcium carbonate, precipitated from our saliva and other fluids, which kills off the bacterial cells comprising the biofilm. Our body's assault on these bacteria appears to be a naturally occurring and regulated process, meaning its part of our immune system. But it's what happens next to the biofilm that I found most intriguing as I instigated my own investigations of tooth decay. These dead bacteria comprising the plaque become encrusted along with the mineral deposits from the saliva on the outer surface of the tooth. This hardened layer is called "dental calculus" or "tartar." I deduced that this outcome is perfectly natural, meaning nature wants it there. But why? It seemed obvious to me: To help protect and strengthen the tooth structure. Still, I also had to ask myself, then why would the dentist be hell-bent on grinding it all off? And not to forget the myriad "plague killers" plied in every supermarket, pharmacy, and dollar store. Surely, not to weaken the tooth?

The answer appears to be the industry's interpretation of how certain bacteria comprising the plaque are responsible for causing tooth decay. Their perception is backed by considerable supportive research and the obvious epidemic of tooth decay flowing through millions of dental chairs worldwide. Thus, dental plaque, no matter what its purported value as a tooth fortifier according to troublemakers like me, *must go*. The alternative they proclaim, while pointing directly at your own teeth, is tooth decay. They are really making me look bad now. But I learned long ago, do not argue with nature.

Well, they can point all day, and stand behind their science all they want, but the fundamental question of why nature creates a tooth fortifier that they feel compelled to remove, isn't really being answered. My opinion from listening to dentists and reading related research abstracts is that dental plaque is not viewed at all as a desirable natural fortifier. It is simply as a "garbage dump" full of acid dripping micro-monsters that must be removed regularly and disposed of. Just like the garbage you put out weekly for the city collector to pick up and get rid of. Now you're really getting nervous, because the suggestion is I'm wanting you to have a mouthful of garbage. Time for some push back — let's look at this plaque business from the natural dental care perspective.

Dental researchers have elucidated that S. *mutans* are naturally aggressive plaque builders. S. *sanguinis*, we recall, are also part of the plaque ecology, but their role seems given to acting on the environment in such a way as to mediate colonization by the mutan streptococci strains. From this, I deduced that *S. mutans*, on the bright side, are great plaque and tartar builders, work horses if you will, to build strong teeth. But if they're so great at it, why temper their populations with *S. sanguinis*? I thought, a stretch at the time, maybe they're an important catalyst, like whatever causes cement to form when you add water to the dry mix. Bear in mind, at this point in my research, I was trying to sift "vitality" from a mouthful of "pathology" to figure out what the heck was going on in our oral cavities. Not exactly a comforting epiphany to tag *S. mutans* as a champion of plaque, *S. sanguinis* a cement catalyst, and their product tartar a blessing of a remorseful Tooth Fairy when the entire dental industry is waging war against all of them! At any rate, everything up to this point seemed "good and natural" to me. But, I knew it was no time to celebrate because there were some obvious holes (as in decay) in my "good and natural" dental model.

Our hero plaque builder, *S. mutans* turns out to be a double-agent, smiling in our face while stabbing us in the back. Notorious for their sugar appetites, they also metabolize sucrose (common table sugar) into lactic acid. Lactic acid can burn a hole in a tooth just like Coca-Cola dissolved that nail. Oh, oh! Further, this acid can become trapped behind the biofilm as dental calculus forms. Tooth structure is then degraded by this acid, and nearby gingiva (our "gums") may be irritated too, as well as other periodontal tissues and structures that support the teeth.[1] What is worse, these offending and greedy bacteria actually flourish in the acidic environment, and are able to proliferate "out of control," creating more and more lactic acid and, consequently, a low pH microbiome that destroys otherwise symbiotic bacteria. Oh, no! So, as more acid erodes its way in, destroying more and more tooth structure ("pathophysiology of tooth decay"), nerves within the tooth's dermal root "pulp" sense trouble coming, and sound the alarm. This message is transported to our brains via the central nervous system in the form of pain.[1] Off to the dentist we go! Who then tells us that there are other bone and tooth lesions, as well.

Unfortunately, there is still more to the gloomy picture. In fact, tooth decay is only a very limited, if not myopic view, of a much bigger problem. That's right, we're back to the Whole Body Inflammatory Disease (aka "WHID"). As our unnatural diet — which includes anything we put in our bodies, including industrialized foods, drugs, and chemicals — reaches our stomach and lower intestines, digestive bacteria also begin to wage war

---

[1]*Periodontium*, collectively the supporting structures of the teeth including the cementum, the periodontal ligament, the bone of the alveolar process, and the gums. Readers who have suffered from inflammations and infections of these structures have certainly been briefed on them by their periodontists — dental "specialists" who treat them. But, in an attempt to bring those readers comfort, we don't really need to know anything more about them to navigate the path ahead towards healing them.

against each other. It is this intestinal battleground that really turbo charges the war in the teeth we are losing. We all know the feeling of "acid indigestion," which is why the drug counters at Wal-Mart are full of acid reducers and proton-pump inhibitors (Omeprazole), big sellers in the pharmaceutical side of the industrialized food chain. Acid loving intestinal bacteria thrive with our carbohydrate "enriched" diets too, and like S. mutans, get right down to the business of dominating once symbiotic bacteria also inhabiting the bowel. The lactic acid produced there erodes (ulcerates) the mucosa of the intestines, and contamination (endotoxins and exotoxins: dead and living bacteria, cellular waste, and other debris) is then absorbed into the vascular system and transported everywhere from brain to toe nail. Nerve receptors in the "confused" stomach now release (hydrochloric) acid, sending us more "red alert" information in the form of pain.[1]

It's my opinion that this digestive contamination acts upon enzymes and bacterial populations throughout our body, with the pernicious effect of breaking it down — including our teeth. Not surprisingly, oral mucosal permeability facilitates added contamination to the cesspool.[2] According to Dr. Mark Burhenne, one of the most vocal and forward thinking dentists I've crossed paths with to date:

> The intestinal microbiome determines so much of our health and well-being — these microbes in our gut have been linked to controlling mental health, weight, and even our resistance to dementia. But there's more to the story. The opening to this intestinal tract — the oral microbiome — has long been ignored . . . the mouth is an essential part of a healthy gut. We abuse the precious and delicate environment in our mouths; antibacterial mouthwashes wipe out microbial diversity, which in turn affects the microbiome in our gut and the health of the rest of our bodies. In my 30 years of practice as a dentist, I've seen many people who have fallen through the cracks due to our healthcare system's failure to understand the mouth-body connection. As a dentist, I've learned that to help my patients get healthy, we can't look at just the mouth in isolation — we have to see the whole picture.[3]

Indeed, scientists have detected S. mutans colonized in heart valve tissue and arterial plaque.[4] Depending on the severity of this systemic toxicity, our immune system may break down completely, like a failed dam, and virulent microbes once in symbiotic balance, are now free to unleash a torrent of destruction anywhere across our bodies, *and I*

---

[1]The progression can also occur insidiously without pain. And to further complicate the picture, the presence of pain may be attributed to matters other than decay. There's a bit to sort through here, but it can wait until later chapters when we get deeper into nuance of tooth structure and its immune system.

[2]Evidence abounds that oral contamination may also gain entrance to the bloodstream whenever opportunity presents (e.g., dental cleanings and surgeries). Harsh flossing causing bleeding of the gums would be another contributor.

[3]https://askthedentist.com/oral-microbiome/

[4]Nakano, K; Inaba, H; Nomura, R; Nemoto, H; Takeda, M; Yoshioka, H; Matsue, H; Takahashi, T; et al. (2006). *Detection of cariogenic Streptococcus mutans in extirpated heart valve and atheromatous plaque specimens. Journal of Clinical Microbiology.* 44 (9): 3313-7.]

(*Above, facing page*) Lovely Indian ladies from the Brazilian Amazon with healthy teeth, free of decay — the result of a healthy lifestyle and diet not contaminated by our industrialized food chain. We don't have to go into the jungles of South America to solve our epidemic of oral disease. We just need to respect the "healing powers of nature" and adjust our lifestyles accordingly.

mean anywhere. Vital organs may fail. Skin, hair, and nails visibly corrupt as we see in diabetics. Otherwise normal metabolic processes serving our vitality and healthy cellular growth, now degrade or mutate into weird, cancerous growth any place where blood flows and these virulent microbes can take hold. Sugar metabolism, once manageable, now devolves into insulin resistance (IR), giving rise to unprecedented levels of Type II diabetes. Neither are brain cells spared, and, not surprisingly, we also see today a range of neuromuscular disorders, cancer (choose your type!), Multiple Sclerosis, Muscular Dystrophy, and Alzheimer's Syndrome. I'm not imagining all of this, you can't turn on the TV for 15 minutes and not hear about it from Pharma pitching their drugs, and countless, highly suspicious non-profits trying to get in on the cash flow — not to mention very litigious attorneys running class-action lawsuits. Arguably what's happening here on TV is a direct corollary of what's happening to our body's vast colonies of microbes out of balance and struggling to survive.

Chapter summary

Now, the foregoing is quite the indictment of what we are doing to our bod-

ies, and broadly so as a species through the international industrialization of our food chain — and by "food chain," I mean anything we put into our bodies, from diet, to Pharma, to the polluted air we breathe. In this interpretation, we can't just look to our oral cavity for solutions to tooth decay. Indeed, we're clearly faced with the eerie and more than likely possibility that what's going on down in our bowels is also contributing to oral disease. In effect, we've opened a two way microbial highway — between the mouth and intestines — acting "behind the scenes" as the principal driving force sustaining oral disease. Which is to say, we can brush our teeth "'til Kingdom Comes" and they're still going to rot if we keep that highway open. What is more, because we're really talking about WHID, it's a super contaminated highway that leads to all parts of our bodies. It's a highway we need to shut down. And our natural dental care program is going to do just that by drawing on our own body's healing mechanisms. Indeed, we are reminded of an important and comforting admonition of the Hippocratic Oath: *Respect the healing powers of nature.* But so far as we are concerned, what exactly does the term "healing powers" mean, and how do these powers apply to our troubled teeth?

Chapter Four

How Nature Maintains and Heals Our Teeth and Gums

I will begin this chapter with perhaps the most outrageous idea that is going to hit the pages of this book. If you can survive it, and are willing to work your way emotionally through and past it to the exonerating premise that concludes this chapter, then you're probably one of those who will benefit from the natural dental care program I'm going to present later in this book. But I've got to say it, and say it now, even though in my "mind's ear," I can hear the stampede of running feet of the feint-hearted crushing past each other to get into the nearest dental office, crying and clamoring to get back in as a "reformed" traitorous patient, who's learned their lesson in misguided ventures into the wanton "edge city" world of purported low life dental quackery espoused by yours truly. Here goes:

> *Dental decay is as natural as the setting sun and has nothing to do with pathology! It's suppose to happen, we are to let it happen, we are to embrace it, we are to facilitate it, and most important of all, we are to stop worrying about it.*

Hello? Is anybody still here?

There is, of course, a calculated hidden oxymoron in the statement: *nature uses dental decay to build strong teeth.* But, if I had added that to the statement, for sure, no one, and I mean no one, would be here to read another word.

Continuing . . .

If it is true, as some pretty good science has revealed, that nature didn't fumble the ball in selecting us with good teeth, witness our ancient Paleolithic ancestors. Yet nature created a plaque scaffold that harbor harmful microbiota that can and do rot our teeth. Logically, something must be amiss, if nowhere else than in our minds. One would think that this contradiction would particularly ring true in the minds of the dental faithful who diligently brush their teeth 3 times a day, use recommended toothpastes and dental rinses, eat healthfully and exercise, see the dentist once or twice annually for check-ups and plaque removal – and still have cavities, get crowns, then root canals, and now wear dentures or posts or bare gums. But the fact is, this Paleo dental connection to our past is given no berth for discussion in the dentist's chair. Caveman style natural dental care is deigned completely irrelevant. Patients truly have no way of knowing about such a thing, and talk about leaving the chair would only strike terror in them anyway. Unless, of course, some weird epiphany leads them to the Internet or bookstore to search out books like mine, and aren't put off by the maligning by the dental industry that a book like this is traitorous, disingenuous, and dangerous - good for nothing and at best a poor substitute for Paleo fossil fuels to heat your home with. It is interesting that in writing this book the only persons willing to discuss any of this with me, were some ladies who have lost nearly all of their teeth and are wearing dentures. Nothing to lose, I suppose, so why not? Others I've approached to talk about this literally throw their hands up in the air, rotating them left and right like a referee signaling a TKO in a boxing match, "I'm doing just fine, and I'm not going to talk with you about my teeth." Meaning: "How dare you get personal with me." Ok, ok. I get it, you don't want me in your mouth, and, really I don't want in it either.

The "right to lose" one's teeth to decay is embedded in the travesty of modern dentistry itself. People roundly believe in the 3 Tooth Truths BS and dare not cross the dentist feeding it to them. And they'll be damned if they're going to be called out to an alternative viewpoint, invested as they are in the acceptance of dental decay, fillings, root canals, etc. A friend's young son, facing a series of root canals, panicked at the suggestion of talking with me, and actually took a tone of anger — a reaction far afield from the usual suspiciousness expressed through rolling eyes. Ironically, unawareness of the underlying paradox — the mechanism of tooth decay being central to building strong, healthy teeth — is actually fabricated unwittingly by dentists themselves and their patients. Let me explain.

Working together under the premise that nature has failed all of humanity, neither dentist nor patient actually gives a single thought as to how nature *might* be able to heal and maintain healthy teeth. The fact that teeth continue to decay over time is proof in their camp that nothing is actually getting healed, and, therefore, there's nothing healed to maintain. Maintenance is simply the time between dental treatments, or more to the point, until more tooth mass is going to be removed. For the patient, it is either a time of wishful hopefulness, or submission to dread if the dentist has prognosticated trouble ahead. Any such discussion suggesting "natural alternatives" would be heretical to the partnership anyway, in fact, tantamount to a threat: To the provider's income and to the sanctity of one's troubled oral cavity. To outsiders, like myself, our focus naturally turns to the unthinkable prospect that what nature intentionally puts asunder, nature intentionally puts back together again. Nature, we assume, did not endow us with Humpty Dumpty teeth.

What I am saying, and that is purportedly so unthinkable, is that *tooth decay is a perfectly normal, natural, and necessary process*. But one nature has also embedded with reparative mechanisms. The two are meant to go together. In other words, we cannot have one without the other. Although counterintuitive, there are analogous systems in nature that also occur in our bodies. An example is our fingernails. For the nail to grow continuously down for us to clip away (just like the horse's hoof[1]), it somehow must be released by the body. It does this through the mechanism of proteolysis: a specialized class of enzymes that proliferate naturally in the dermal bed below the nail (made of keratin, a type of protein), attack and digest ("lyse") the normal cellular bonding structures. So disposed, new growth is able to move towards the end of the finger, lysing, re-bonding, lysing, re-bonding, etc., all along the way. Nature "maintains" by "destroying" cellular adhesion mechanisms, then repairs the "breakdown" by generating new bonds. In effect, it is a continuous process of "shedding" old growth. Similar processes allow us to shed old skin and hair. The horse's hoof sheds this way too, creating an entirely new hoof in approximately 9 months.

But, while all this is good and fine, our industrialized food chain, dental science, dentists and their patients, have colluded unwittingly to cause the natural processes to go awry. The antidote to this collusion, in my opinion, is for the patient to break loose from the dental chair for a moment and start thinking about what's *not* happening in terms of nature's healing mechanisms. What's so hard about doing just that for starters? Pushback from the dentists using the 3 Tooth Truth weaponry is the main reason. The dentist will tell you, "Once decay has commenced, you have to remove it or it will continue to progress and you will eventually lose the tooth; further [we are reminded *ad nauseam*], you won't be able to grow new replacement teeth once they're lost." This edict given to every dental patient after the Tooth Fairy is long gone, is now tantamount to the law of the land. So much so that governments could easily legislate and promulgate it without an iota of public resistance. But profound fears of tooth decay and losing one's teeth is so indelibly ingrained in the minds of modern humans, there is no need for such legislation. The masses will just do as they're told, as they do now.

---

[1] I'll visit this connection between tooth and hoof again as we look closer at very similar pathologies that effect both. But I need to lay more ground work first. Patience!

And that means getting, and staying, in the chair. And not looking to nature for answers.

The fact is, nature can and does heal our teeth, and does so 24/7. It does this by providing us with tooth DNA that is equipped with healing mechanisms passed down through the millennia from our ancient Paleolithic antecedents. Where they got this, you'll have to ask God or your local university evolutionary biologist. How it does this, however, is what should be of interest to us. Unfortunately, there just aren't any flashing neon signs and billboards along the freeway heading towards the dentist's office announcing the fact: "Detour: Go Here To See How Nature Heals & Maintains Healthy Teeth!"

The biggest part of the dental conundrum rests in not recognizing that dental healing isn't conditional exclusively to the microbiology of the oral cavity. As discussed in the previous chapter — and this is of paramount importance to nature's oral healing mechanisms — it's equally beholden to our entire digestive microbiology, which impacts every cell comprising our bodies. Thus, viewed from a genuine holistic perspective, oral and intestinal health share a common matrix of healing mechanisms that encompass both degradation and restoration. Nature has set this up to work in our favor through balanced microbiotic communities. We can either respect and facilitate balance, or we can break it down. Nature will not let us play one colony against the other just to treat symptoms; indeed, if we feed the body in one place, we feed it everywhere else. Logically, feeding it in a healthful way, we get one result, feeding it in an unhealthful way, we facilitate a contraposing result of disease pathophysiology. It's up to us, therefore, to do the right thing, or suffer the consequences of weakening and breaking down our natural healing mechanisms. Which is to say, our immune system.

The significance of this biological system of balances is without parallel, in my opinion. And no where is this more evident than in the pages of any mainstream medical or dental text on the treatment of disease. Because these disciplines, entranced by high technology and hell-bent on treating symptoms, seem only capable of mustering the very, very thinnest veneer of holistic care. What little there is, is marginalized and institutionalized as "preventive care," which really only works if you're eating healthfully, exercising, and staying out of their clutches — like hospitals and dental chairs. But, again, we are faced with the fact that from their perspectives, there's no money in being healthy. So prevention stays tucked away in the corner virtually out of sight, seen as irrelevant and remote from Hippocrates's admonition to "respect the healing powers of nature," as the vegetarian's cookbook is to the cannibal. So, yes, we're back once again to Whole Body Inflammatory Disease (WHID)! And back once more to Cooper:

> We brush our teeth and we floss, and we think that we've got good oral hygiene. But [we're] completely failing to deal with the underlying problem. Ten years from now, I think we're going to find that the whole microbiome is a key part of what you get monitored for and treated for.

His point being, by ignoring the underlying cause – colonies of bacteria aggressively overwhelming others — any healing at the tooth, which will require cooperative behavior among the microbes, is doomed to failure. Aided, of course, by the dentist's whirring drill and our negligence fostered by our unquestioning obedience to the 3 Tooth Truths. So, how does natural healing work? And what path do we take to get around all the anti-holistic roadblocks set up by the dental industry to keep us in the chair?

First, we are wise to arm ourselves with an understanding of how the tooth heals itself, or at least tries to, and, as Cooper has foretold, come to realize how the microbiome needs to be "fed" so they will get along without killing each other off, including our teeth. Then we can talk about how we're going to intervene to bring "peace and harmony" and have healthy teeth. Which is to say, if we don't understand causality (pathogenesis) of oral disease and the countering mechanisms of healing on some level, we'll never be able to withstand the dental industry's terroristic pushback. We'll find ourselves once again strapped to the dentist's chair, our teeth indelibly branded by the 3 Tooth Truths with crowns and amalgams and posts and dentures, begging the dentist to drill away as repentance for flirting with the ideas of fringe dental pseudoscience and conspiracy theories.

Microbiome

First, we *must* recognize that many strains of bacteria occupy just about every niche in our bodies, and by the gazillions. They are a part of us. Or, maybe we are a part of them? *Second*, we respect that nature intends for them to be there, but in balanced communities. "Balance" is the key point we must intellectually anchor ourselves to. But clearly, we know our microbiome of bacteria is in trouble and anything but "balanced" when we, as a species, aren't either – witness collectively our rotting teeth, cancer, and other metabolic disorders characterized by cellular chaos and human suffering (also true with horses![1]). According to Cooper (and other scientists), bacteria make up approximately 90 percent of the cells comprising our bodies. Researchers have now shown that these cells actually communicate with each other, including with our brain cells! (Does that mean we're missing their message?) Since they're a part of us, we actually need to treat the "whole" lot of them with more respect! Like Cooper, I believe that we focus too much on "I" and not enough on the microbiome ("them") that spans our entire body, including our teeth. So, let's think "us," not "me," for starters.

The good news is that by feeding the growing biofilm in our teeth in healthier ways, their extended family elsewhere in our bodies, including in our bowels, will benefit too, and, vice versa. Which is to say, if we please one, we'll be pleasing everyone, including "I." One big happy family! Conversely, if dietarily we're disrupting our oral biofilm and getting tooth decay and gingivitis, we're certainly distressing bacteria colonies elsewhere in our bodies. Because what passes through our oral cavity will eventually reach our bowels to be processed, and like a boomerang, be delivered right back to haunt us throughout our bodies via our vascular system. One big unhappy family. *Sigh.* Diet, by default, then becomes our main reliable ally in the effort to restore a balanced microbiome of warring bacteria broadly across our bodies, including those in our mouths.

Our digestive biology

It now seems certain that, while overt symptoms of trouble in the oral cavity are obvious to us due to decay and pain, insidious problems may go unnoticed in the gastrointestinal tract that processes our food and nutrient intake. What comes to mind immediately are the consequences of widespread use of acid-reducing meds. As an example, using proton-pump inhibitors (chemicals that neutralize copious amounts of excess gastric acid where our bodies produce it) like Omeprazole can suppress the painful symptoms of acid indigestion fairly effectively in the stomach, which is good, necessary and certainly advised to avoid ulcers.[2] But, here

---

[1]*Laminitis: An Equine Plague* J. Jackson. Several chapters provide numerous examples of WHID in equines.
[2]This recommendation comes with the caveat that stomach acid is also part of normal digestion. Too little acid has been implicated in other digestive disorders, but excessive acid production carries its own destructive path. In either case, supporting research becomes circular, in my opinion, most of it ignores WHID, and, more importantly, ignores holistic healing of our digestive microbiota. Thus, Omeprazole seems the safer path to take within the holistic context of our natural dental care program, while science attempts to unravel its contradictions fostered by treating symptoms instead of causality.

again, we are only treating the obvious *symptoms* of inflammation. Which seduces us to not think about causality. Namely, a microbiome that just isn't being fed right and is warning us of impending deeper, more insidious problems if we just keep putting off the inevitable. So we can't just stop there at the symptoms threshold, we must move forward to tackle causality. But how, in the defense of our mushy molars and cracking canines, are we to go about this?

First, by recognizing that within this broader digestive interpretation, dental interventions deployed to remove decayed tooth mass harboring *S. mutan* colonies multiplying out of control and dumping acid everywhere in our teeth and gums, are tantamount to using Omeprazole to quiet acid "burn" up-and-down the gastric tract. As I tell my own clients with horses suffering from deadly laminitis, "To heal your horse's feet, you must feed all of your horse's bacteria, beginning with the vital digestive bacteria." Well, what's good for the horse is good for us, too. Otherwise, we are back to "treating symptoms" again rather than causality, and the vast, replicating colonies of *S. mutans* and other pernicious, acid loving bacteria elsewhere in our bodies, will reward you with all the tooth decay you could possibly hope for. And, for many people with weakened immune systems, they're looking at more cancer of the bowels and elsewhere in their bodies. Nature is no fool and knows we are playing Russian roulette with our bodies in a "fool's paradise." As I've written in one of my books, "Nature will not tolerate incompetence."

Strategically, a prebiotic diet (more on that later) will have to play an important role in our tooth healing if we are to calm our ubiquitous microbial colonies that are hell bent on creating acid everywhere. Fortunately, we don't have to be fanatical foodists to feed our biofilm the right way to accomplish this. I will take diet up in Chapter 6, where I hope to bring "moderation" to the dinner table, where "Paleo" and "Veggie" eccentrics are in open battle to debunk, defend or invade each others' turfs, not at all unlike uncooperative bacteria waging war on each other in our decaying teeth and elsewhere in our bodies, just to get what they think they deserve to eat. You and I aren't the only ones at the dinner table.

Healing pathways "off the grid" of conventional dental care

What all this equates to by way of "healing" is relatively simple: what we eat, what we do in the way of dental hygiene, and how we exercise (Chapter 8) — will determine the outcome in our teeth. Nature will take care of the rest. We don't have to know all about tooth structure, which is infinitely complex (typical of nature) and seemingly beyond modern dentistry in the field to understand itself. Nor do we need to try to understand the scientific elucidations of dental pathophysiology, as a lot of it is clearly jaded by corporate driven outcomes. Anymore than we have to travel to archeological dig sites to inspect Paleolithic skulls to see great teeth. That's not our job and it isn't necessary. Now, that's good news and that's where we're heading!

So, if all the colonies (by species) of bacteria are meant to be there, then it's really a question of microbial balance. Here's where natural dental hygiene and natural diet have to come together. Briefly, we need to let the mutans streptococci do their job in laying the plaque-tartar barrier to strengthen the outer layer of our teeth. That's the role nature created for them, so let's let them have at it. And to do their job well, we must feed them what they need and deserve, or they will turn on us. For the mutans streptococci, this means sugar. And either we give it up to them, or it's war. And we must stop trying to kill them off. This really peeves them, and I don't blame them either. Now, I know this is super counterintuitive to the 3 Tooth Truths ingrained in us, and an affront to the dental industry who will declare war against us to defend their status quo, but nature's a pretty powerful ally that wants to side with us on both battle fronts.

Our role, divined by nature, is to regulate the mutans streptococci population so they don't get the upper hand over competing colonies and cause too much decay while building the tartar defense. So, to restore and sustain balance, we also have to put some controls on their voracious appetites by feeding other mediating bacterial colonies with whom they are cohabitating. We'll get into all of that in Chapter 7. Once again, it is a startling oxymoron that our teeth must undergo constant decay through bacterial acidosis in order to possess a stronger structure. But that's the way nature wants it. It worked for our earliest ancestors, and it will work for us. So don't panic! *Be strong, be brave, be informed!* We don't want to smell the dentist's drill burring your teeth into smoky "tooth dust" in the air!

§

Before leaving this chapter, I feel compelled go a bit deeper into this misguided clash between dental pathophysiology and contraposing healing forces brought on by the dental industry. It's not been the easiest task in writing this part of the book, but tracking the progression of dental disease does yield the dividend of understanding where some of nature's immune defenses are strategically positioned. We need to know that they are there and are struggling in the face of wide-spread neglect by misinformed patients who are under assault by the dental industry. Knowing they are there will give us the encouragement to take them seriously and provide them with the unprecedented support they will require from us. It is here, too, that antagonists from the threatened dental industry will, and not with a simple dismissive flick of the hand in thin air, launch their most vicious attacks yet against the foundations of natural dental care. *Be brave, be strong, be informed!*

Pathogenesis of dental caries

Causality of tooth decay is conventionally attributed by the modern dental industry to two related but separate pathways, both of which we are going to challenge[1]:

1. Biofilm (dental plaque) attaching to the teeth and maturing to become cariogenic (causing decay) environments is due to bacteria in the biofilm that produce acid in the presence of fermentable carbohydrates such as sucrose, fructose, and glucose.

2. People from the lower end of the socioeconomic scale are more susceptible to dental caries than people from the upper end of the socioeconomic scale.

My opinion is that #1 must be refuted because it ignores the biological fact that plaque is necessary to build strong tartar structure. #2 ignores our species DNA, which researchers like Cooper have shown that the richest and the poorest among us have the same chance at having great teeth; but because everyone is feeding off the same industrialized food chain, everyone is suffering. Really, the only difference is that the poor can't afford the high end, cosmetic dental procedures, and the rich can. But, other than the contrived hype of Hollywood actors in TV ads proclaiming "breakthroughs" in dental implants, tooth dyes, and so forth, no one, and I mean no one, is dying to jump up and announce the miserable facts of the matter to the world. Dental sufferers, rich or poor, are as closeted as they come.

What is "hidden" in the penumbra of these shallow and misleading edicts is that "nature and facts" speak otherwise: #1 and #2, like the 3 Tooth Truths, are only true in the confines of an industrialized society whose dental industry, like a hungry T-Rex pursuing its prey, must consume your teeth if it is to survive. The amount of money entrenched in the dental industry from tools and equipment, to

---

[1]Selwitz R. H.; Ismail A. I.; Pitts N. B. (2007). "Dental caries". *Lancet.* **369** (9555): 51–59.

office space, to staff salaries and training, to research, to attorneys on both sides, drugs, on and on, is staggering. This ravenous monster isn't' about to let go of your teeth in its current incarnation. And patients are too intimidated and ill-informed by the 3 Tooth Truths to stand up for themselves.

Pathophysiology

The following very brief discussion is about as far as I'm willing to go into the deeper realms of tooth disease. It is a morbid environment of darkness and professional violence against nature that I'd rather not think about, to tell you the truth. Nor, on a technical level, is it necessary to understand by the more lay reader. But some readers, more academically oriented, may find it of interest. I see this type of "war with nature" in my own profession from my perspective as a holistic horse care practitioner. Nevertheless, I think something needs to be said, if not definitively, at least as a way to point a finger towards a better direction based on nature's immune defense network. If nothing is said, then the message won't be heard at all.

So, first, let's be clear where we stand on the pathogenesis (origin) and pathophysiology (functional changes) of dental disease. A problematic diet that includes anything we put into our bodies, including (excess) sugar, toxic drugs, chemicals, vaccinations, and a host of other toxic substances nature never intended to get inside us, is what sets off dental disease — and WHID, to be precise. Current science states, and of this much I am in agreement, our teeth are bathed in rich minerals derived from fluids secreted into our mouths in the form of saliva. This causes a mineralization, or surface restoration, of the hard tissues (e.g., enamel) of the tooth that are eroded by acidification caused by the actions of mutan streptococci colonies in the biofilm. Where I depart from current science is on this point: while I agree that when the rate of remineralization is slower than the acidification rate occurring in the biofilm — resulting in

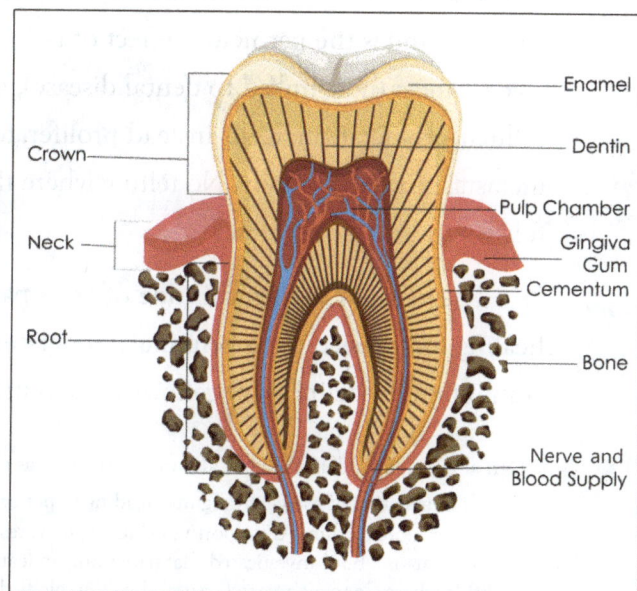

Enamel

Dentin

Crown

Pulp Chamber

Gingiva
Gum

Neck

Cementum

Root

Bone

Nerve and
Blood Supply

lesions on the hard tooth structures — the process itself is, nevertheless, perfectly natural and necessary, as I've explained earlier, *to build strong teeth.*

Dental scientists may argue that we are both talking about "balances versus imbalances" in the biofilm, but my point of departure is that the intellectual and technical emphasis needs to shift towards "what is natural" rather than "what is pathogenic" in the oral cavity, and, moreover, must also be inclusive of the "whole body." Because what dental science has defaulted to, or at least translated to as dental practices in the chair, is to ignore this and simply "search and destroy" anything and everything that hints of pathology in their path — including the tartar barrier, demonized as "plaque."[1] Tooth structures are then left "wide open" for catastrophic assault in the form of partial and total mass removal, toxic fluoridated toothpastes and dental rinses, and — in the research phase as I write this — vaccinations to wipe out *S. mutans* and *S. sobrinus* species as though they were polio, and manipulation of DNA in critical enzymes essential for normal mutans metabolism of sugar (sucrose).

For fearful patients, this may sound good at the surface (of their teeth), but we are tinkering here with the healing mechanisms nature put in place, and in so doing, asking for trouble at the core of what it means to be human (or horse!). Careful, careful! We are not just talking about compromising tooth strength by eliminating tartar. Because our many strains of bacteria are also commensal with each other, vital digestive mechanisms are also implicated through their symbiotic relationships, thus, we are impacting vital cellular growth. What comes immediately to mind is the pernicious effect of radiation and chemical therapy in treating colon cancer (now linked to dental disease), which destroys or suppresses normal cellular proliferation. Cells instead proliferate pathologically as tumors capable of metastasizing systemically. No telling where the "two way highway" will then lead to!

The panacea to turn this tide of gross pathology is to understand that nature's healing forces need to be genuinely understood and facilitated – not ignored or committed to the industry gallows to be outright destroyed as is the practice right

---

[1]Dental researchers identify tartar, also called calculus, as excessive calcified plaque (biofilm) accumulation resulting in irritation of the gingiva (gums) leading to periodontitis and gum disease. And, not surprisingly, a platform for more biofilm! As supportive evidence, their research abstract points to pathogenic calculus in "ancient humans." But I investigated that paper only to learn that ancient meant medieval Europe! Hardly paleolithic where Cooper's research reveals healthy calcified plaque (tartar). Further it corruption ignores the impact of WHID.

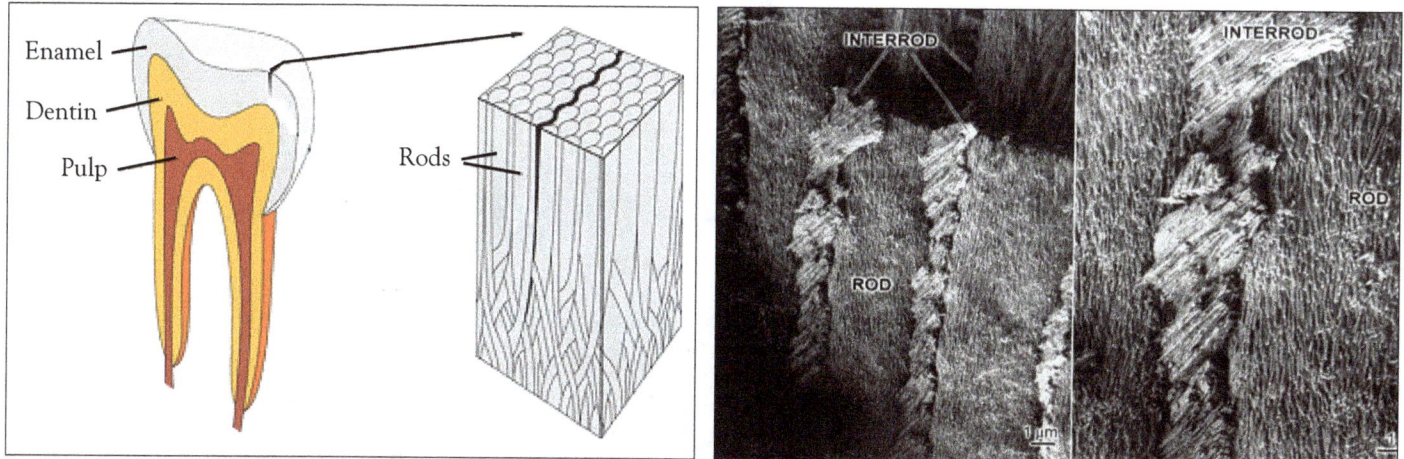

now, and increasingly so. In my own organization of practitioners, we have a saying, "Ignore all pathology" and, instead, embrace holistically the healing powers of nature to attenuate the contraposing forces of disease. Good advice! Otherwise one is too vulnerable and easily swept up from intimidation by the dental industry's fear-mongering pitchmen with the consequence of being strapped back in that chair tighter than ever.

Scientists studying cariogenic progression breeching the hard (calcified) tooth structures (enamel, dentin, and cementum) and entering the dermal (pulp) structures within are aware on some level of nature's self-defense (healing) mechanisms along the way, but, from what I've researched, understanding is limited and muddled in the largely uncontested and accepted models of combating oral disease. For example, as offending bacteria degrade the enamel's "rod" structures (*above*), cavitation occurs in the direction of the rods until the outer dentin body is penetrated. As offensive bacteria metabolize sugar into lactic acid, dentin is penetrated and demineralized. This opens the door for new waves of acid-loving bacteria, and, ultimately, more acidification followed by more cavitation. Because complex cells called ameloblasts naturally degrade and depart from the enamel structure following completion of amelogenesis (formation of enamel), new enamel cannot be created. Or so it has been assumed. New research that I will point to later, however, suggests otherwise. The enamel breech can be repaired using saliva (tartar), compensatory gels and compounds, and stem cell therapy, thereby precluding the need for invasive dentistry. This information is laid out in Chapters 6, 9 and my Epilogue.

Going deeper into tooth structure, cavitation is aided destructively in the dentin structure by pathologically proliferating proteolytic enzymes. The core of the

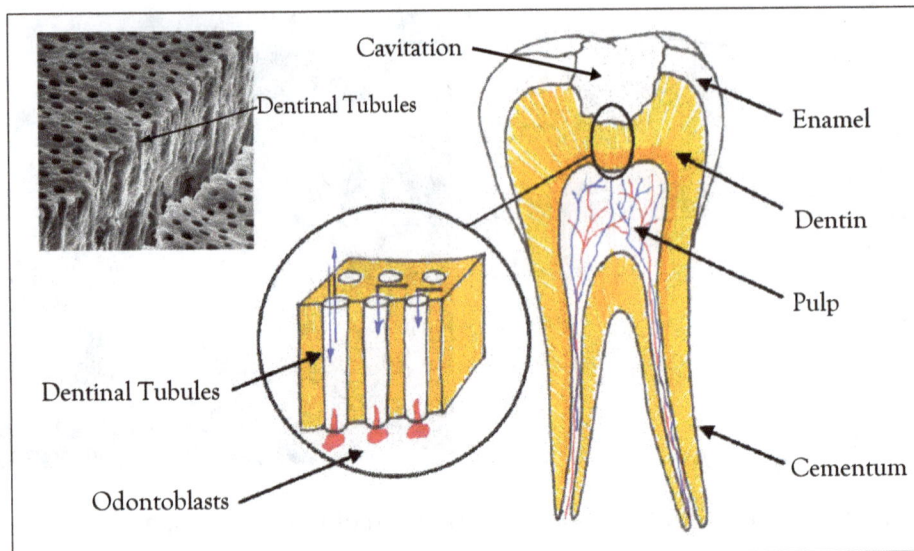

tooth's hard structure is now truly imperiled, and the presumption here, too, is that more penetrating restoration — crown hardware and maybe more — will be needed to save the tooth. Further, if the pulp chamber is also breeched and infected then we are talking the likelihood of losing the tooth altogether. Natural dental care suggests another strategy altogether . . .

We recognize that proliferating daughter cells of the extracellular matrix responsible for new dentin at its deepest layers along the pulp barrier scaffold are selected by nature to initiate biologic repair responses. Fluids within the tubular structure of the dentin carry antibodies (immunoglobulins) from the immune system to combat infection,[1] aided by mineralization which constricts the tubules to obstruct bacterial migration.[2] Other reparative dentin is manufactered by specialized cells comprising the pulp, stimulated into action by growth factors; here, the effect is to deform tubular structure to force a misalignment with caries progression in an attempt to further obstruct bacterial migration.[3] A similar invasive process occurs at the gum line or "neck" of the tooth root where calcified cementum integrates with the dentin. Nothing to ignore! This is nature's attempt to occlude the bacterial pathway progressing towards the inner layer of dermis (pulp chamber). Our immune system fighting hard for us, but in a losing battle — unless we take holistic countermeasures!

Holistically, enzymatic degradation is tempered and restored through diet

[1]Summit, James B., J. William Robbins, and Richard S. Schwartz. "Fundamentals of Operative Dentistry: A Contemporary Approach." 2nd edition. Carol Stream, Illinois, Quintessence Publishing Co, Inc, 2001, p. 13.
[2]Nanci, A. *Ten Cate's Oral Histology.* 2013, p. 166.
[3]Summit, Ibid. p. 14.

(which is what I do with horses suffering from WHID). Infection is controlled by stimulating populations of patrolling M2-macrophages (white blood cells) that are selected by nature to engulf and digest contamination in the pulp environment so that tissue can repair and heal; this is facilitated principally through diet and exercise. These healing mechanisms are nature's last line of defense on our behalf, and we can aid them through holistic, non-invasive interventions that do not favor unbalanced microbial populations within us. The admonition, "Ignore all Pathology," equates intervention with prevention, thus, the holistic interventions of natural dental care are a double-edged sword in the balancing of our microbiota communities.

Here, I have to add that I have long questioned the industry's position that cavitation results solely from aciduric mutans streptococci colonies that have breeched the enamel barrier. In other words, the pathological migration of cavitation progresses only from the outside of the tooth, and advances from there *inward* towards the pulp (dermal) chamber. But, we've seen [Nakano, *et al*], S. *mutan* colonies have been implicated in the pathogenesis of diseased heart valves as well as atheromatous (plaque) debris within degraded fatty tissue of inner arterial walls elsewhere in the body.[1] It is, therefore, conceivable that migrating S. *mutan* or its toxic plaque debris is entering the tooth matrix through the vascular "backdoor" of the dermis feeding the pulp chamber, thereby stimulating proteolysis of the dentin structure from within, or more likely, from both directions if the enamel barrier has also been breeched. Pollitt *et al*, cited earlier, found S. *bovis* and S. *equinus* debris – implicated in the pathogenesis of laminitis in horses due to saturation of Fructan (sugar) lactosis in the hind gut – within the dermal-epidermal lamellar scaffold of extirpated biopsied hoof samples. How did it get there? No doubt, the same way S. *Mutans* did. I wonder, what harmful opportunistic bacterial strains, and viruses known to attack and insert themselves in bacteria, are colonizing "outposts" (sanctuaries) for themselves in organs far from their native habitats elsewhere in our bodies? My natural dental care program assumes the worst, and we will build immune system defenses everywhere we can in our bodies through holistic intervention. I've taken exactly this approach in my healing of horses and I see no reason why it doesn't apply here. As I wrote in my introduction, "healing patterns in nature repeat themselves broadly across many species."

[1] Thought to have breached the oral mucosa and migrated via the vascular system.

Chapter summary

At this point, it goes without saying, many of our mouths have already been invaded and "repaired" by the dentist who has the technology to detect the effects of the bacterial invasion and internal proteolytic destruction not seeable with our own eyes. Further, the previously hypersensitive nerve matrix may shut down completely while the carious progression continues along in a more insidious manner unbeknownst to us. Bolstered by the 3 Tooth Truths, dentists and their assistants — and not to forget the patient strapped to the chair — have no reason whatsoever to give further examination or credence to nature's subtle healing mechanisms still at work deep within the tooth, but which are clearly overwhelmed and struggling.

I've looked at the research done here and it is clear that there are forward thinking scientists, and some dentists (very few), who are looking at this, and who actually see the "hidden picture" of nature's healing field in the background. And, in a glimmer of hope, not all dentists agree with or approve at what's going on in their industry's chairs. In fact, there are dentists who have come to realize that "preserving" mass is of paramount importance and that reckless, profit-driven procedures to take mass at all cost are tantamount to criminal malpractice. But what do we do in the realm of natural dental care? My dialogues with mainstream dentists, who put people in the chair daily, are revealing of utter disdain towards any notion of "natural healing," although, in my opinion, they are actually pretty naïve about what this means. Even more bizarre is their almost contemptuous attitudes towards stem cell research (discussed in the Epilogue). Regardless, it is clear both from common sense and genuinely good science, that nature has positioned healing mechanisms all the way down the pathology corridor, from our saliva, to the diverse biofilm colonies on the enamel, to the tooth's root, to the vascular system, to our bowels, to what we are eating. It is all there.[1]

So it comes down to us — victims of "old school" dentistry — to help, if not prod, forward thinking dentists and research scientists to pioneer a new way. But, really, doesn't it come down to us to grab the reins and take control of our own bodies, too? After all, none of our healing mechanisms are going to do us, including our teeth, any good if we just stand idly by in fear and denial, letting, if not begging, the industry to just have their way with us with our silent complicity.

---

[1] The parallels with my own work dealing holistically with deep pathology at the horse's hoof (and body) are astonishingly similar.

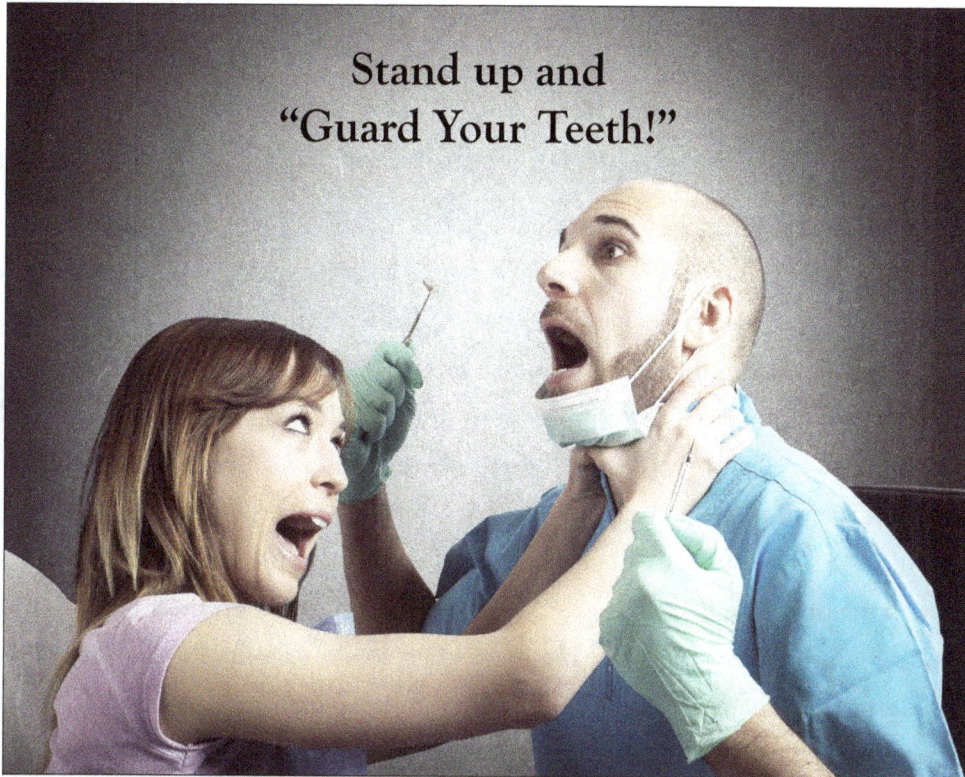

Stand up and
"Guard Your Teeth!"

You'd think we were under a gag order. Potential healing forces will remain in disarray and "out of balance" in what the industry's defenders will humor as a contrived and delusional "New Age Magical Mystery Zone" and denigrate it accordingly as hyperbolic extremism at its very worst.

Before I tell you what I do, let's look a little closer at why traditional dental care is really all about stopgap measures and not healing. The industry is engaged in a continuous losing battle against the progression of dental disease that needs "victims" if it is to survive in its current form as a profitable industry. It's really all about Big $ and sustaining an industry that's not a whole lot different operationally than the Neolithic "dentists" with their stopgap and ineffective bow drills and beeswax. The difference is in technology and cunning advertising designed to alternately terrorize and seduce the public into acceptance of eternal chairdom — not serve what should be the technical objective of genuine healing and prevention. We're talking about a serious addiction that unrelentingly and insatiably intends only to mechanically modify and remove tooth mass. Enough is enough!

Chapter Five

Guard Your Teeth!
Why Traditional Dental Care Fails Us

The principal problem with modern dentistry as I've come to understand it, is the presumption that both dental calculus and decaying tooth mass must be removed. And, further, that "bad" bacteria must be eradicated and removed too — along with any collateral bacteria that, like an innocent person killed in a random drive-by shooting, happen to be in the path of their whirring drills and plaque killing dental rinses. Failure to do so, they sternly and impatiently warn, will result in the loss of the tooth itself, more teeth in time, and possible periodontal disease. Our entire oral cavity, they will go so far as to say, is imperiled. This apocalyptic prophesying of dental doomsday seems logical, and, more often than not — if not always — is exactly what happens. However, their logic and outcomes are predicated entirely on the faulty premise of their 3 Tooth Truths and correspondent regimens of dental interventions that rig their results. As willing participants, our teeth don't have a chance. This is the "self-fulfilling prophecy" of dental religiosos, which explains precisely why traditional dental care fails us no differently than the waxed Neolithic dental bow did in its day. Our natural dental care perspective is quite different.

Removal of dental calculus

Grinding and rinsing away nature's intended veneer of calculus (tartar) not only deprives our teeth of added structural protection, it upsets, if not destroys altogether, the diverse bacterial population of the biofilm that continuously manufactures healthy dental calculus. This devastation results in perpetual immature colonies of surviving bacteria which favor the mutans streptococci clans, even more so if the diet isn't right. Given the upper hand, these plaque building bacteria are free to injure the exposed tooth structure and burrow their way in with lactic acid. Further, researchers have found that debridement of plaque by dental technicians and bruising or penetration of the nearby epidermal-dermal mucosal barrier by the dentist's drill and probes can export rampaging S. Mutans and their toxic wastes from the oral cavity via the vascular system to remote parts of the body, including, it is my opinion, the root matrix of the tooth. This dispersion only compounds the same problem emanating from proliferating harmful digestive bacteria and other opportunistic microbes, including viruses, in the lower intestines. In other words, we're back again to Whole Body Inflammatory Disease (WHID). Hence, we are talking about the systematic and catastrophic importation of disease triggers into our bodies that break down our immune systems.

Our natural dental care response to the dental industry's *antiplaque syndrome* is straightforward: plaque, like biofilm from which it is derived, is a natural constituent of the tooth. In short, let's create healthy plaque and tartar, and leave it alone!

Removal of tooth mass

Removing dental plaque (biofilm) carries its own toll on the oral cavity (and elsewhere), but so does removal of underlying tooth mass, including tartar. Generally, this occurs therapeutically with discovery by the dentist of carious lesions on the surface of the tooth or more progressed cavitation detected by radiographs but not yet detected via the tooth's innervations felt as pain by the patient. Even more extensive loss of tooth structure results from exigent procedures when the patient arrives with advanced decay and acute pain as the disease has progressed in close proximity of the innervated root matrix. If too far gone, then mass extraction is extended to include the root mass as well! But in all cases, nature's mechanisms to supply healthful healing nutrients to the tooth's dermis are of no real concern to the dental industry. Not because the dentist drilling away does not care, they are simply dismissed because the 3 Tooth Truths are proof that they are ineffectual.

While dentists are aware of the mechanisms for epidermal remineralization of the decayed mass via salivation, they truly do not understand the holistic measures necessary to bring biological order and balance to the ecology of our vast inner biospheres, including the oral cavity. Thus, the institutionalized sequence of "decay > filling > crown > root canal > total extraction > denture/implant" is virtually guaranteed. And, we cannot forget, whether for the sake of oral function or in the name of dental cosmetics, this is where the big money is to be made — not in tooth vitality.

And here I wish to highlight a new understanding: *cavities are not a disease of the tooth itself, or of the oral cavity*, rather they are a symptom of WHID. Diabetes, Cushing's, cancer, multiple sclerosis, muscular dystrophy, and other immune system disorders - it's the same thing, all are also symptoms of this digestive disorder, not isolated regional diseases somewhere on or in our bodies that just "arrived out of nowhere." This statement, of course, constitutes medical heresy from the stand-

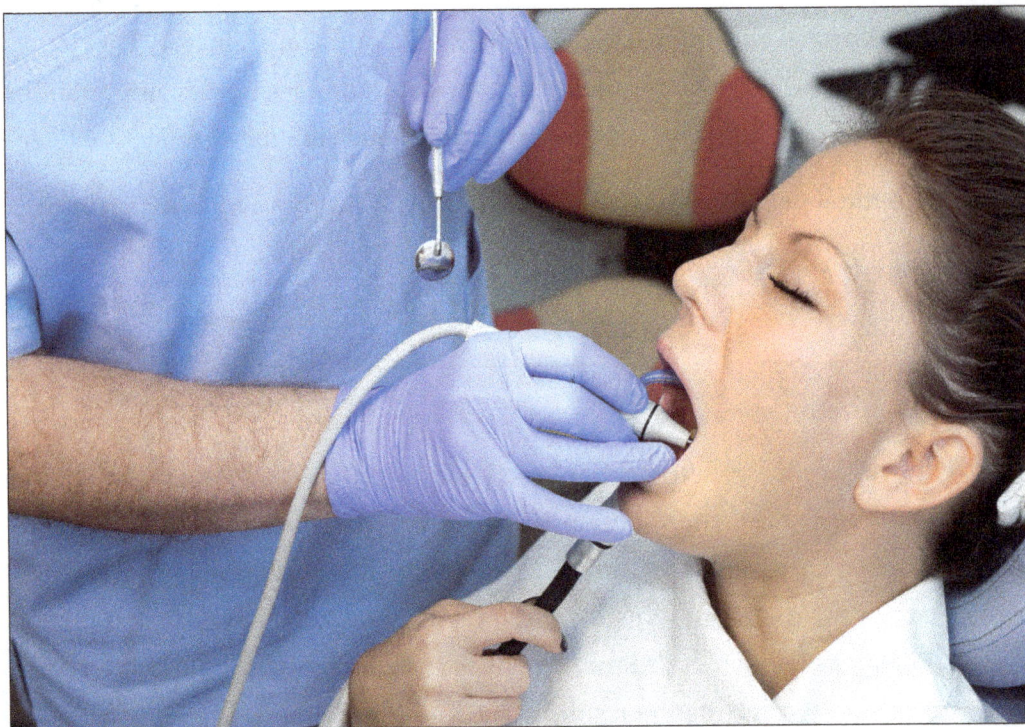

point of mainstream medicine and dentistry. We expect this from the mega-industries that profiteer in the billions of dollars annually from misdiagnosed metabolic syndromes rightfully classified as WHID.

But, with all due respect to our detractors, the science of natural dental care is not "fringe." It is rooted largely in credible mainstream science with effective, common sense interventions carried out by people themselves. From this perspective, the only counter discussion I want to raise here concerns the pernicious effects of traditional dental care on nature's "healing powers." Their premise, in the simplest of terms, is, biologically bogus and catastrophic to our teeth. It is all about war: pitting profit-driven science and technology against nature. In contrast, the natural approach to dental care intervenes before mass is removed and ramps up nature's healing mechanisms through reasonable dietary, exercise and oral hygiene regimes. This is the subject of Chapter 6 and the main focus of this book you have been anticipating. So, with the foundational information in position, let's go there right now to put our teeth back together the way nature intended.

§

Summary

The problem with going to our dentists is that they are completely armed up in "mind and technology" to remove tooth mass. Our tooth mass! They can't help themselves. It's like an addiction for them. In fact, it is an addiction. If you study the dentist's behavior while talking semi-pleasantly with them about "alternatives," their hand wringing begins amid more sweaty 3 Tooth Truth rhetoric. Even anger and denial emerge — the exact behavior of a narcotics addict when confronted with their addiction. Nor can you get tough with them outside the shifting concentric boundaries of their contrived "3 Tooth Truth" fables they continue to manipulate and aim at us in their own defense. If you try to push them outside these emotional envelops, they might even tell you to get out and not come back. I've riled a few myself. It's very traumatic for them, so best to leave them alone and stay away. That's my advice. But if you're one to learn the hard way, go ahead and give it a go. I'll be waiting for you right here when you've had enough of their BS.

Chapter Six

Natural Dental Care for You and Your Teeth!

I hope the reader didn't just jump impulsively to this chapter without reading everything that came before. You'll need a clear understanding of the previous information to make the most of the practical information laid out here. In fact, I recommend you put the book away and go back to whatever you were doing to your teeth in that dental chair if you think you're going to just brush your teeth in some new way and that's all there's to it. That's because this chapter isn't just about cleaning your teeth in new ways, it's about supporting your whole body's efforts to stop the progression of disease in your mouth (and body), and to prevent it from happening in the future. As I hope I've made it perfectly clear in previous chapters, oral disease simply cannot be stopped in your mouth alone, which is why tooth decay continues to be rampant worldwide. Nevertheless, what we're going to do in our mouths is very important and probably very different from what you're accustomed to — and for most people — I would say critically so.

§

We recall that the dental industry's treatment of oral disease is really treating "symptoms," not the disease itself, whose primary source of trouble lies elsewhere in your body. That's "Whole Body Inflammatory Disease (WHID)," in case you've forgotten. We don't want to keep making the same mistakes. Do nothing but go after the symptoms in your mouth, and I can virtually guarantee you failure. I will send you right back to Chapter 1 to start over if you're going to be guilty of the charge. I don't want you to give "natural dental care" a bad name because you think you know better.

This chapter's model for natural dental care approaches matters holistically: from oral hygiene, to nutrition, to exercise. That's right, exercise! If you think in your spare time you're going to just lay around on the sofa watching TV, sitting at your computer reading fake news, lounging around down at the senior center playing checkers, or being depressed in bed — and just brush your teeth, whine, and create worry warts, you've got news coming. But it's actually good news! There's nothing here that you can't do! It's basically simple and rational — do it yourself stuff. And for many people, most important of all - it is very affordable. Certainly relative to traditional dental care. And I imagine your dental bills are probably mounting as you read this, or you're worried because you can't even afford a dentist. In the course of doing my research for this book, including drawing on my personal observations from the past in life, I am very empathetic with the many people I've known or witnessed with devastated teeth and gums who had few or no financial resources to do anything about it. Particularly those struggling below the poverty line and who are too proud to use our government's "entitlement" programs, or have been turned back by dentists because you won't be able to afford what they want to do to you. I contrived an experiment to test this concern of mine. Here's what happened:

I randomly selected and scheduled an appointment with a dentist to have my teeth evaluated and taken care of. Support staff up front were leery of me from the start because I said I had no dental insurance, but would pay for the service out of my own pocket. I managed to make it to the chair, where I told the dentist I was an Army veteran (true) but didn't really have much money to pay for anything (a lie). Knowing how a genuine professional "probe" is supposed to take place in the chair, I wasn't surprised to discover that no real probe even took place as it was over in less than a minute, patted on the back, "everything's fine," and shown the door. No x-rays, no plaque removal, no tooth dictation to staff, nothing. I suspect his eyes were probably closed too when the mirror-less probe was underway. The message was clear, however: no money, no service. Service, of course, I had no genuine interest in, but it confirmed in my mind that "profit motive" was central to this dentist, and probably many others, too. But there was something else that bothered me about this guy, a smile with a hint of snarky superciliousness to be sure as he headed me out of the examination room. I'm sure not all dentists out there are like this, but we know for sure there's at least one. Another thing for sure, being an army veteran wasn't any help either! So, if you are one of the many who lost all or most of your teeth and are now wearing dentures because you could not afford professional dental care, I regret this natural dental care program wasn't around in time to make a positive difference in your life, and at very little cost.

Natural dental care encompasses cleaning our own teeth, building a natural dental calculus (tartar), and using nutrition (probiotic diet) to enhance healing mechanisms deeper within the tooth structure that can only be reached via the vascular system. For the most part, this entails feeding healthful, balanced colonies of oral and digestive microbes (Chapter 7) — with an end to "search and destroy" missions to take them out! — and just enough exercise to optimize circulation to get vital nutrients where they need to go (Chapter 8). Exercise will also build opti-mal muscle mass that will feed off the healthy fats we will need to eat to restore vitality. None of this will require trips to the dentist, unless they are interested in helping, but it will require you to make some trips to your bathroom sink to clean your teeth, the grocery store for some healing foods, the Internet to get a few supplies and ingredients you probably won't find in your local stores (yet!), and a very small workout space in your house or apartment. Right out front, we're not talking about dieting, eating weird stuff, or becoming gym rats. But it is about us taking some unique personal responsibilities if we are to escape the clutches of a ravenous, money-hungry dental industry that, in its current form, must excavate our tooth mass if it is to survive. This is combat, and we need sturdy defenses against what's going to come at us from the industry's entrenched and well-oiled propaganda machine to scare you back into that chair. And re-member, dentists have cavities like the rest of us; so, if it brings you some com-fort, trust that there will be dentists reading all of this right along with you, "just in case."

So, let's proceed with that understanding, and with my understanding that you've studied (not just perused) previous chapters closely and know what I mean when I say that the epicenter of dental disease is that "two way highway." It's your teeth that are at stake, and being honest with oneself is as crucial as are the neces-sary changes in our behavior centered on natural healing and natural dental care. I would add that if you are a victim of other metabolic disorders, what follows just may help you with those too. But first up for discussion are some things we don't want to use and do to our teeth. This is important information because it represents our first significant departure from the type of traditional dental care most of us have been doing our entire lives. And with all that put to rest, we'll get into its replacement: natural dental care!

Two dental no-no's!

Conventionally, oral hygiene has meant getting rid of food stuck to and be-tween our teeth and in our gums by brushing and flossing. And also removing dental plaque (most effectively at the dentist's office), and using Fluoride tooth-pastes and dental rinses sanctioned by mainstream dental associations to kill off bacteria and plaque (biofilm). We'll be making a number of "left and right turns" away from convention here, but if you've read the previous chapters it will all make sense (I hope). At any rate, what follows is what I do because it works and it is simple and it is inexpensive. My only regret is that I didn't start doing it earlier in my life when I still possessed more of my teeth. But then, the information we now have available to us to process and act on is relatively new. And without the availability of this information made possible by the Internet, there's no doubt I never would have climbed out of the dentist's chair myself to write this book. The industry's counter arguments are deeply ingrained, intimidating, and backed by government, establishment science, manufacturers, the pharmaceutical industry, and the dentist staring you in the face with drill in hand! But we'll deal with all of that by being informed.

1. ADA approved and "alternative" toothpastes

The principal problem with commercial tooth pastes and powders notably sanctioned by the American Dental Association, which we find on supermarket and drugstore shelves everywhere, is that they are essentially useless when it comes to balancing the oral microbiota. In fact, they aren't intended to balance anything. Quite the contrary, whatever their purported efficacy, they are formulated to do one thing: destroy bacteria, plaque (biofilm), and tartar. Certainly not to bring them into harmonious coexistence!

Virtually all are laced with artificial sweeteners (and other chemicals) that don't belong in the mouths of any living thing. Before continuing, let's address these types of "man-made" sweeteners and why I oppose their use. The dental industry, government, and many researchers (but not all of

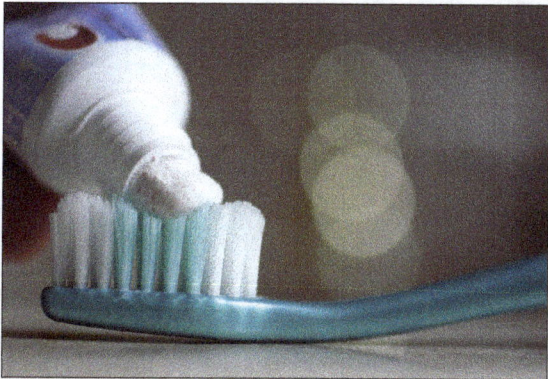

Crest Pro-Health Clean Cinnamon, 0.454% stannous fluoride, 0.16% w/v fluoride ion.

them[1]), recommend them as "safe" sugar substitutes. All are fairly unanimous that "intense sweeteners" (non-caloric) like aspartame, saccharin, and sulfame, and "bulk sweeteners" (caloric) like sorbitol, xylitol, and mannitol, are not metabolized to acids by oral microorganisms, and, therefore, are not cariogenic. Further, they point to the sugar alcohols (e.g., xylitol) that are effectively cariostatic (inhibit caries). So, they argue, they are good ingredients to put in toothpastes to make the pastes more palatable ("sweet") to people. What they're doing is playing to our sugar addiction without the calories. By drawing our sensory perceptions to "sweetness," which all humans have a love affair with, and with the added assurance that the artificial sugar means no calories or cavities, we're "sold" on sweetened toothpaste.

But the problem here is that it is not known how theses ersatz sugars — with the possible exception of xylitol (discussed in Chapter 9) — and other chemicals they are formulated with interact with many other microorganisms in our digestive system, including those remote from the oral cavity via WHID's two way highway. We recall this was a concern of Dr. Mark Burhenne (p. 27) and other dental experts,[1] and it certainly is an overriding concern of mine treating WHID in laminitic horses.[2] The fact that dental caries persists in spite of widespread use of toothpastes with artificial sweeteners supports my thesis that the dental industry is simply ignoring the root cause of microbiotic imbalances in our digestive system due to unnatural dietary habits coupled to the assault on tartar fortification. One group of dental scientists has even said as much:

> Albeit sugar is associated with the dental diseases like dental caries, we emphasize the fact that sugar alone is not the sole determinant of these diseases. To prevent dental diseases, oral health care workers should persuade their patients to adopt special dietary programs and educate patients and motivate them to alter their customary dietary behavior. Furthermore, such health education must compete with the food manufacturers marketing techniques to significantly reduce dental caries in the population at large.[3]

This is truly a clarion call to elevate the significance of a "reasonably natural diet" as part of the human diet to deal front and center with tooth decay. And it certainly is with our natural dental program. At any rate, whether in our toothpastes or foods, we are not putting "real food" or "natural ingredients" in our bodies, and our digestive microbiota know this and will respond as necessary to protect themselves, or opportunistically expand their colonies to such degrees as to become pathological troublemakers themselves like the mutans streptococci. More nutritious and far safer, in my opinion, is that we use organic, naturally occurring sugars (e.g., honey and fresh fruits) strategically and in moderation so as not to in-

---

[1]Worth reading from Harvard University with warnings on sugar substitutes: http://www.health.harvard.edu/blog/artificial-sweeteners-sugar-free-but-at-what-cost-201207165030

[2]*Laminitis: A Plague.* pp. 8-10.

[3]P. Gupta, N. Gupta, A. P. Pawar, S. S. Birajdar, A. S. Natt, and H. P. Singh. *Role of Sugar and Sugar Substitutes in Dental Caries: A Review.* (US National Library of Medicine National Institutes of Health: https://www.ncbi.nlm.nih.gov/pmc/articles/PMC3893787/)

vite dangerous acidic shifts in our body chemistry that favor WHID.

§

Toothpastes also typically contain detergents (e.g., Sodium lauryl sulfate), fluorides (e.g., stannous fluoride), and abrasives (silica). These are the "big 3" plaque killers. Increasingly, "natural" toothpaste manufacturers are offering substitutes for the first two ingredients, contesting the science offered by some researchers that they are beneficial to the oral cavity. But in keeping with the anti-plaque fear syndrome, all manufacturers deploy abrasives such as silica (sand), baking soda, calcium carbonate, hydrated alumina, and dicalcium phosphates for plaque removal with the toothbrush in your hand providing the grinding force — until you return to the dentist assistant's chair to really get the job done.

My point in this discussion of modern toothpastes is that they are all search-and-destroy chemical compounds aimed, purportedly, at defending and remineralizing tooth enamel. But the fact remains, it is at the expensive of nature's natural protective tartar layer. If the efficacy of toothpastes were true as they claim, and everyone is clearly using the stuff, then dental caries would be a thing of the past. Which isn't the case at all. More important is that our Paleolithic ancestors used none of it and had great teeth. And that is worth contemplating in formulating any genuine natural dental care regime.

2. ADA approved and "natural" dental rinses

Next up to go are the infamous "dental rinses" your dentist recommends as the "2nd Tier" assault weapon on our teeth. And manufacturers — conventional and alternative alike — are up for delivering it big time. If you own any, it's high time to pour them down the drain, or better, take them to the toxic waste pickup station run by your city or county solid waste department and dispose of them there. There are many versions on the supermarket shelves, and most of them carry the same seal of approval of the dental industry you see on toothpastes, "When used in an approved program of [blah, blah, blah . . . .]." So, you should know better already then to use them. Just look at the list of chemicals and artificial (and free) sugars in them, masked behind a plethora of fruity flavors to choose from. Join us renegades pitching them into the toxic waste bin at the recycling center. But, just to confirm in your own mind, before christening their exodus from your life, take a closer at the label and you'll read, "Kills germs that cause bad breath and tooth decay!" And, not surprisingly, "Rids teeth of unhealthy plaque." That's good 'ol "search and destroy" rhetoric again. I have to admit, the dental industry's galaxy of germ killing, plaque destroying, and tooth mass removing technology is one of the most brutal money-making schemes in the annals of civilization. And all those billions of cavities worldwide to justify it! What in heaven's name would our Paleolithic ancestors with great teeth have thought of it all as they ran for their caves?

Armed up with fluoride and alcohol!

Natural Dental Care to the Rescue!

Our natural dental care program, at face, appears to involve doing the same exact things we've all done (or are doing) conventionally: namely, we brush and we rinse. An important difference is 1) How and with what type of brush; 2) What we're brushing with in lieu of toothpaste; 3) What we're rinsing with instead of the germ and plaque killers; and 4) how we combine brushing and rinsing. This is a bit of a "stepwise" process, but we'll take it on step by step.

Having said this, I have to throw this reminder in once again — the strategic importance of diet and exercise. Because of the 3 Tooth Truth mythology, these are concerns both dentists and many lazy patients will just shake their heads and roll their eyes at. I mean, when's the last time you sat down with your dentist and talked extensively about diet and exercise and their relationship to healthy teeth? I can't think of any time. Yet, these are so integrally important, words can't describe. Naturally, I've given them their due in their own following chapters.

Okay, moving forward, let's get down to this business of brushing and rinsing as you'll want to get going on both fronts right away, particularly if your teeth and gums are in trouble.

Brushing

I had no idea how people brushed their teeth until I was thrown in with a ton of guys in the Army years ago (Jan/1968 – Jan/1970). During basic training, we all had to live together in close quarters for longer than I care to remember. I saw it all. What I've extrapolated from that life experience, and recent interviews with people about their dental habits as part of writing this book, is that most people probably brush their teeth either too little or too much and harshly so, according the ADA recommendations. With the exception of people suffering acute or chronic tooth pain, those who do "too little" is because their teeth aren't hurting. That kind of makes sense. Why bother to go to great lengths, minute after minute, brushing away for some indeterminate amount of time if nothing seems to be wrong? Of course, such behavior obviously leads to a mouthful of problems. But this is no time for such indolence if you're guilty as charged, as a little more effort than the least possible isn't where we want to go in our heads and mouths.

At the other extreme, the tooth brushing brutes among us want no part of this wimpy, lackadaisical indifference. To them, this is war driven by 3 Tooth Truth terrorism and the fear of not getting rid of all those decay causing germs and dental plaque that are going to put them back in the dentist's chair. Why, if I were to put a can of Ajax in their hands, they would probably deploy that too, along with a bottle of Crystal Draino fortified with spinning metal corkscrews to really scour out the entire oral cavity! In the end, alas, both "wimp" and "brute" suffer the same fate because their brushing isn't going to do any good anyway, like the other 3.5 billion victims worldwide. But, let's look at matters from the perspective of the bacteria, first in the days of our Paleolithic ancestors, afterwards in the present.

The Paleolithic "bad" bacteria (and, recall, they aren't "bad guys" after all) would not have even been remotely intimidated by brushing, which wasn't happening anyway. Through specialized evolutionary adaptations they were more than ready for such a fight. Their sticky plaque had long glued them to the Paleolith's healthy tooth enamel that no theoretical tooth brush could breech. Unfortunately, salivation had sent the whole lot of these bacterial marauders to an early grave, transforming them into an armored wall of calculus (tartar). Not that they cared, because new generations of acid dumping bacteria (*S. mutans* and *sobrinus*), who had long been devouring natural sugars sourced and eaten by our ancient ancestors, were just as content to keep themselves attached to the tartar. Of course, salivation did them in too, and every future generation since until Paleoliths became farming Neoliths, and the new "civilized" diets and early dentistry began opening the tartar barrier to the greatest and longest lasting bacterial infestation of *H. sapiens* in our specie's entire history. Which brings us to the gloomy present.

With the tartar barrier soon tagged as the bad guy, and search and destroy missions sanctioned everywhere by the ADA and their international counterparts,

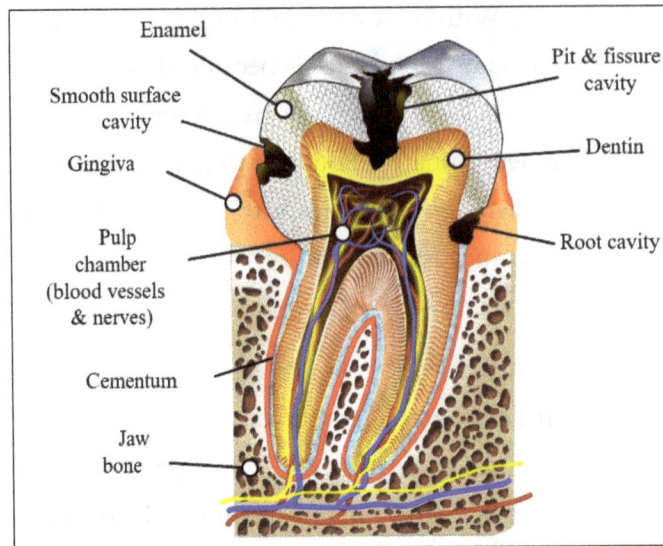

the microbial invaders found many ways to take out our teeth (image above), aided of course by Truth Tooth Mythology and the Two Dental No-No's. But matters are actually much worse than we think. *S. mutans* is only one of many oral Streptococci species inhabiting the oral cavity, and apparently different species seem to target various parts of the tooth structure. Sort of like Paleolithic tribes staking out different hunting and gathering territories. While not all are harmful in their natural state, the shifting biology of the oral cavity due to WHID and their commensal relationship to *S. mutans* appears to convert others into pathogenic invaders too. Cohabiting yeasts and fungi have also been implicated in the amplification of *S. mutan* cariogenesis, also due to WHID. Many of these aggregate symbiotic relationships would otherwise be instrumental in creating healthy tartar in the Paleolithic model. Sooner or later, their giant free for all turns into a literal nerve wracking experience for you. You are left with a frayed toothbrush and a toothache. Boom! Off to the dentist you go, sobbing and kneeling and begging forgiveness before your dentist, who's probably got a mouth full of dentures himself and wants you to join him so he feels better about himself. Maybe so you can commiserate in Tooth Truth misery together. Have you ever seen a dentist with great teeth? The wives of two dentists I interviewed confided their husbands' greatest fear in life is to go before another dentist! What does that tell you?

Unlike the wimp who has given up, and the brute — who is still on his futile "search and destroy" mission against the bugs and plaque, and any food that's in the way — and the anal retentives in both camps who are going to cosmetically whiten their teeth with dyeing agents to really cover-up the crime scene, we have no such interest. We want no part of Tooth Truth mythology and whiny, crybaby commiseration. And let me say from the outset, the task of cleaning is actually a relatively minor part of the natural dental care program. We're going in an entirely different direction than what the experts have advised us during our youthful phase of complete naiveté: *to restore biological order.* The important thing to understand in going forward is that we're not at war with what's in our mouths, and we shouldn't be using toothbrushes like weapons of mass destruction anyway! Put 'em down for a moment, along with the toothpastes and dental rinses, and read on!

In terms of actual "brushing," we will be targeting both the gums and teeth with coordinated irrigation. If you're going to use a toothbrush, get one that has soft bristles. The natural food or health stores usually have them, and the bristles are typically made of naturally occurring materials. So, think of the tooth brush as a device for lightly massaging and stimulating the gums and lightly brushing the teeth of any loose (exfoliating) tartar. Instead of worrying about "not removing enough" formative dental plaque, it's time to relax and try to enjoy the experience! What we're going to remove is what nature will let us remove, and what we're leaving is what nature wants us to leave.[1] Nothing more, and nothing less. So, what does that mean exactly?

I mentioned above removing naturally exfoliating tartar. You may be asking yourself, "What is he talking about?" After the biofilm "cures" on your teeth as tartar, it doesn't just keep building up on and on forever. Nature intends this crust to be biodynamic, in other words, to form and shed with natural wear — for example, tartar lost as a consequence of masticating your food, as would have happened with our Paleolithic ancestors. As this exfoliation takes place, new biofilm settles in to replace lost tartar, and a repeating pattern ensues. This brings us to another discussion of salivation.

Salivation, as explained in Chapter 3, is a continuous process that saturates the formative biofilm with remineralizing agents that mediate *S. mutan* (and other microbes discussed earlier) colonization and acidification of the teeth, while simul-

---

[1]In my natural hoof care training program (www.ISNHCP.net), based on a wild horse model (America's mustangs living in the Great Basin) we have the "Four Guiding Principles of the Natural Trim." The first principle states, "Leave that which naturally should be there." The second principle states, "Remove only that which is naturally worn away in the wild."

taneously restoring the tartar barrier. A healthy biofilm platform, therefore, is essential for tartar formation and stability. Our task is to aid the salivation mechanism through exercise, a probiotic diet, and strategic brushing and irrigation of the tartar scaffold. Static tartar fortification isn't possible if we undermine the oral cavity's microbiotic balances with search and destroy dentistry, lousy diets, neglect of oral hygiene, and anti-exercise indolence.

"How much time do I spend brushing and rinsing?"

Brush your teeth when they need it and that will depend on what you've been eating (diet) and when. If your diet is healthful, don't brush right away. Wait an hour or more for digestive enzymes to break down and extract micronutrients from food particles. Your oral cavity will absorb these vital nutrients and deliver them where your body needs them without any conscious input coming from you, other than giving them time to do their work. I was surprised to hear from several persons who say they eat well and have never had dental caries, tell me they have never brushed their teeth. I would have doubted that if it weren't for the fact Paleoliths also never brushed their teeth. But most of us will probably require some oral hygiene if we've been to the dentist and are late catching up on good nutrition.

Brush long enough to reach all the teeth and gums, probably around 30-40 seconds, but no more. Use warm water to dampen the bristles, and then with nothing more than warm water: gargle, rinse, and expectorate. When you are done, consider using a probiotic oral toothpaste, but don't brush it on as it will likely contain abrasives. Instead put a small amount on your index finger and massage it into the gums, and all surfaces of the teeth. If it gets on your tongue, no problem. The paste and massaging will stimulate salivation. Swish the mixture around in your mouth, between your teeth and along the gum lines (called "pulling") then expectorate. Rinse with clean water one or more times and you're done.

But, you are probably wondering, "That's not much time at all." But that's all that's necessary to help exfoliate loose tartar and unformed biofilm debris. Now I know this "brushing brevity" is really going to frustrate or anger the brutes, who are probably all wired-up to detonate the tartar to get it all off. The wimps, in contrast, are relieved by what sounds like do-nothing Oral Pacifism. Jeeeeez, getting these two factions to the center isn't going to be easy! I figure spending that much time brushing my teeth and gums to restore and maintain their health is more than reasonable. The only exception to this timetable is if you are currently experiencing serious toothache issues, and that will require the dentist's intervention. Until you've settled into our natural dental care regimen, and your dental microbiota are colonizing atop your renewed tartar barrier where they belong, instead of on the enamel inside, this will be your course of action.

If you are like many others, either too lazy, too extreme, or too uncertain as how to work the tooth brush in every which way and for how long exactly to be effective, and are unwilling to set a stop watch to go the length, try using a powered irrigating device used for flossing discussed in Chapter 9.

Flossing

Flossing may be a significant contributor to periodontal (gum) disease, right behind WHID. Especially at the hands of the strong-armers who use floss as second-tier weaponry in their assault on the oral "nasties" clinging in between the teeth who won't surrender – namely, dental plaque. But it is actually the dental industry's propaganda that has drawn us into this self-defeating battle line.

Who can forget the magical "red dye" test for plaque – sold over the counter today as "disclosing tablets?" Most of my generation in the U.S. got it in elementary school in the 1950s. I vaguely recall (partially repressed in the unconscious) how I got the dye into my mouth, but once in there, and to my horror (also shared by hundreds of millions of other youngsters) within minutes the truth of my hygienic delinquency was revealed in the mirror: unseen crevices on my teeth and were coated with the red dye. Like Hawthorne's *The Scarlet Letter*, "The Red Dye" was our dental nemesis and badge of dishonor that exposed our hygienic depravity before others. Even though I am color blind to red, I was nevertheless chastised along with everyone else for not floss-

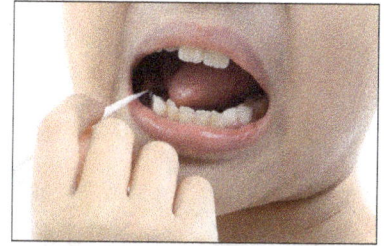

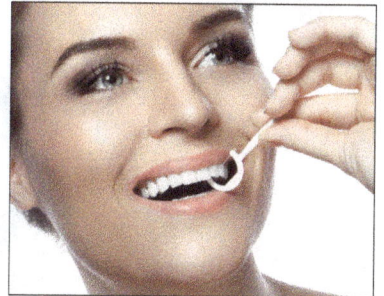

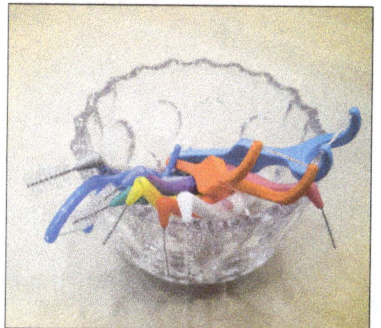

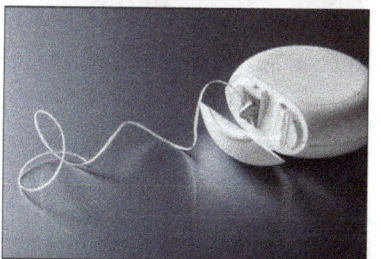

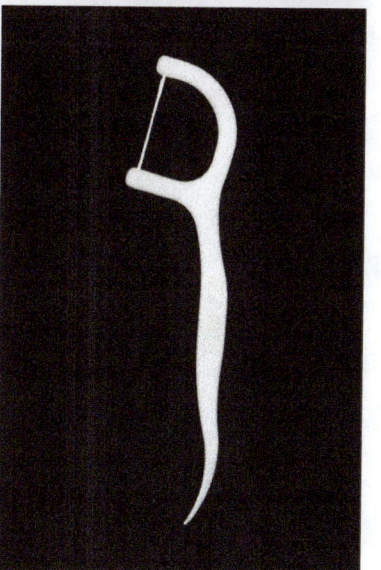

(*Right*) Flossing weaponry.

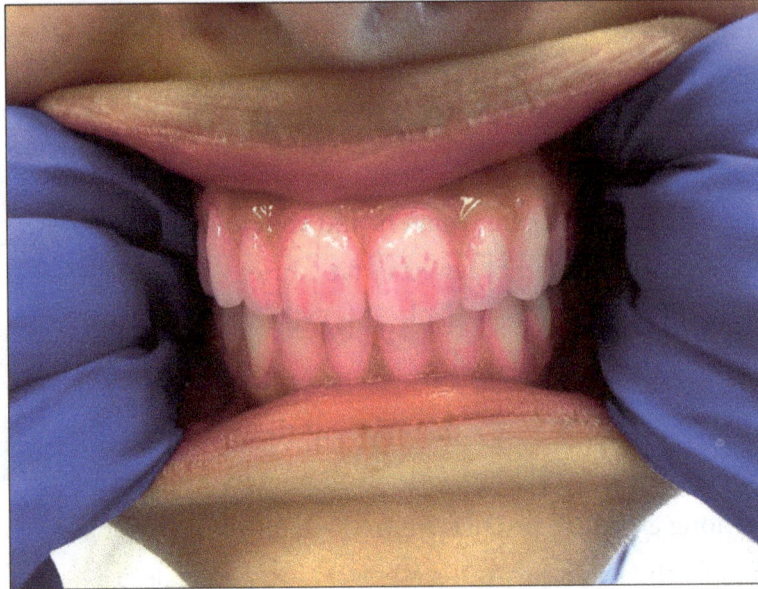

Disclosing tablets are chewable tablets used to make dental plaque visible.

ing enough or at all. Fortunately, my parents were too caught up in their own problems to invest me in floss, although I managed to get a hold of it once on my own like 30 years later.

For the brutes, floss was a godsend, and it was taken into combat as an oral garrote. Most of us complained about the subsequent bleeding, but were assured it would go away, which is true in the absence of gingivitis. Indeed, epidermal structures have their way callusing up when worked beyond the norm. Nevertheless, while I'm not advising anyone to put red dye in their mouths, the presence of red dye on your tooth could be used as evidence of a healthful colony of pre-tartar biofilm. Balancing that colony to "our" benefit, however, is another matter.

So, is flossing advisable? Cautiously so, but not for removing plaque or tartar. If you've got something stuck between your teeth or up in the gingiva (gums), it's probably bugging you and you're going to go after it no matter what anyone says. So go for it, but just enough to get the culprits out that are actually causing you grief. Generally speaking, flossing just between the teeth shouldn't cause harm to the gums, as long as you don't get too brutish about it with them. If you have crowns, you don't want to yank those off – yet! (discussed further in the Epilogue). There are also rinsing technologies that can be used with or without flossing. Most notably are the "water pic" machines that enable you to direct (adjustable) pressurized water at the teeth and gums. My practice has been to use both, but flossing only now, and sparingly, because of its efficacy when used properly.

But, amid this notion of flossing, bear in mind that not everything is meant to come out. In fact, 19th century anthropologists who investigated the teeth of aboriginal peoples around the world that were still untouched or marginally disturbed by civilization, found perfectly healthy teeth with food clung to them that were never brushed nor flossed.[1] We recall that enzymes are naturally present in the oral cavity through Natural Selection. Their role is to break edible things down (lysis) for you. Nature put them there for that purpose. In other words, what's happening between your teeth is part of normal and healthy digestion – feeding your oral cavity with vital nutrients, including your teeth. If nature didn't want space between your teeth to trap food for this healthful reason, we can bet your choppers in any Las Vegas casino that we'd all have selected at the dawn of our species to wear just two teeth: a huge upper and a huge lower, each totaling the entire mass of all our natural teeth combined, plus extra to fill in the gaps! A frightening thought. Thank the Great Spirit for all those food-trapping spaces between our teeth, eh?

Dental healing agents

The natural alternative to conventional toothpastes and dental rinses are their naturopathic[2] counterparts, which I've identified as "healing agents" in our natural dental care program. Each agent serves a different purpose. But combined as they are in this natural dental care program, their synergistic impact is to produce a positive mediation of the oral microbiota colonies.[3] For your reference, here are the healing agents involved and an overview of what they do. I've also included as notations some interesting related research. Specific ingredients are listed with the products in Chapter 9:

- *Remineralizing powder* – Aids your saliva in repairing lesions on the tooth's enamel and in the microcavities behind the calculus (tartar) if they are open through the tartar barrier for whatever reason. An alternative to fluoride compounds.

[1]Weston Price, 1939, *Nutrition and Physical Degeneration*; significantly, Cooper's Paleolithic findings revealed foraged plant debris embedded in tartar.

[2]*Naturopathy*: a system of treatment of disease that avoids drugs and surgery and emphasizes the use of natural agents (e.g., herbs, minerals, and clean water) and physical means such as tissue manipulation (e.g. brushing and gum massage).

[3]*Synergism* — interaction of the healing agents such that the total effect is greater than the sum of the individual effects. Synergy is well-documented by research, although naturopathic medicines ("healing agents") are considered the agents of quackery and pseudoscience by mainstream scientists mentally supersaturated with pathology. I'm accustomed to this arrogance in my own profession and, knowing better, simply ignore the whole lot of them who clearly know less than the naturopaths have forgotten.

- *Natural probiotic foods* – Of strategic importance to the immune system by "feeding" and "balancing" bacteria colonies in the oral cavity and intestines.

- *Transporting liquid agents* – Uptake toxic bacterial cell debris and delivers them to your saliva, which we will then expectorate, followed by the *clean water rinse* (below). Primary ingredients "oxygenate" the vital tissue: naturopathic science holds that enriching the oral cavity with oxygen suppresses (but does not destroy) harmful anaerobic bacteria colonies,[1] thereby aiding the saliva in creating a more natural habitat for a diversity of microbial strains.[2–4]

Note #1: S. *mutans* is a facultative anaerobe, meaning it is capable of switching between aerobic (utilizes oxygen) and anaerobic (requires no oxygen) respiration to survive and function as nature intended. In its anaerobic transformation, S. *mutans* favors lactic acid fermentation. As an aerobic, it manufactures Adenosine troposphere (ATP), a complex organic chemical utilized as energy in many cellular processes.

Note #2: This prebiotic approach is to render a more healthful oral habitat for a diversity of bacterial strains (and other microorganisms). There is reasonable mainstream science heading in this direction, in my opinion, which suggests that S. *mutan* strains and their relationship to dental caries is not exactly cut and dry. How they interact with other microbiota, including other mediating factors in their shared microenvironments, relative to caries is more complex than current "search and destroy" dental dogma would have us all believe. According to one research paper, virulence among S. *mutans* strains is not even consistent, casting doubt on exactly what triggers, or fails to trigger, cariogenic outcomes:

> S. *mutans* is a member of the oral biota, and as such it is in constant interaction with many other microorganisms, as well as with host factors, all likely to play a role in caries pathogenesis. The genetic differences that may contribute to differences in virulence of S. *mutans* strains remain the subject of our research." *Distribution of putative virulence genes in Streptococcus mutans strains does not correlate with caries experience.* [Argimón, S; Caufield, PW (2011). Journal of Clinical Microbiology. 49 (3): 984-92.]

Note #3: Also of interest is the fact that researchers remain speculative as to which bacteria species are the actual players in the cariogenic process and at what stage. Citing over 100 research abstracts, Takahashi, et al., speculate that both mutans streptococci and non-mutans streptococci are involved in the developmental stages of caries:

> Many acidogenic (acid producing) and aciduric (acid tolerant) bacteria are involved in caries. Environmental acidification is the main determinant of the phenotypic and genotypic changes that occur in the microflora during caries. ["The Role of Bacteria in the Caries Process: Ecological Perspectives." *Journal of Dental Research.* Volume: 90 issue: 3, page(s): 294-303. Authors: N. Takahashi, Division of Oral Ecology and Biochemistry, Department of Oral Biology, Tohoku University Graduate School of Dentistry, 4-1 Seiryo-machi, Aoba-ku, Sendai, 980-8575, Japan. B. Nyvad, School of Dentistry, Faculty of Health Sciences, University of Aarhus, Denmark.]

Note #4: Natural selection has been implicated by other researchers, proposing that prehistoric *S. mutans* evolved into the devil that convention identifies it as today under pressure from Neolithic agriculture and the subsequent industrialization of our food chain. According to T. Hoshino et al.:

> "Our findings suggest that the genus *Streptococcus* acquired the gtf genes via horizontal gene transfer. Through the acquisition of GTFs, *Str. mutans* became capable of forming cariogenic dental biofilms. Thus, our data support the idea that the pandemic of dental caries is likely to have been caused by not only anthropological factors but also the evolution of *Str. mutans*." ["Evolution of Cariogenic Character in Streptococcus mutans: Horizontal Transmission of Glycosyl Hydrolase Family 70 Genes." Tomonori Hoshino, Taku Fujiwara, and Shigetada Kawabata. *Scientific Reports*, 2012; 2: 518.]

But this presumes the absence of sugars in the Paleolithic diet, for which there is no evidence. To the contrary, honeybee species (e.g., *Apis mellifera*) were well-distributed before the advent of *H. sapiens sapiens*, and knowing humans as we do, they undoubtedly harvested honey from their hives. Other sources of sugar — fructan rich fruits, for example — were also available to the Paleo diet before the advent of agriculture. I can't imagine that *S. mutans* didn't hunger for sugar any less then than now. More likely, opportunistic symbiosis among mutans and non-mutans streptococci as discussed above takes us closer to the truth, and specifically when humans began to exploit native species of flora that nature never intended.

- *Clean water rinse* – This is significant because water constitutes about 65% of our bodies by volume/weight, more than any other substance, and because it is a mediator of the microhabitats for the many strains of microorganisms in our bodies. We will use it additionally as a "finishing rinse" in our natural dental care program.

Okay, the foregoing is a mouthful to process for sure! The good news is that you don't have to figure out the science nor the ingredients that are involved. They're safe, effective, and it's all been sorted out by people in the burgeoning natural care industries who have been gnawing away at the dental and medical industries for longer than I've been around. My personal product recommendations with specific instructions – and cautions! – are listed in Chapter 9.

Our objective in using these healing agents is to help *naturalize* the habitat of our oral microbiota to bring order and balance among their diverse colonies. This is another significant departure from conventional dental care, which creates a toxic oral environment with its fluoridated toothpastes, mouthwashes, and city water, unleashing an unbridled chemical assault on our oral colonies of bacteria and their nascent biofilm, matured dental plaque, and the food particles between our teeth. Holistic restoration of the disordered oral microbiota is incongruous with the 3 Tooth Truth rationales and interventions of "search and destroy" allopathic dentistry and its allied "over the counter" Pharma. In contrast, the guiding philosophy, nature-based science, and holistic therapeutic interventions (brushing, healing agents, pulling, diet, and exercise) of natural dental care provide a refreshing and badly needed platform for integrated prevention and healing of cariogenic processes, including the underlying pathology of Whole Body Inflammatory Disease (WHID) that must also be brought under control.

In the industrialized mouthful of pathology we've been handed since childhood, the concept of a "balanced community of microbiota" means nothing to drill happy dentists, the institutions that train them, and the manufacturing and Pharma industries that serve them and their brainwashed patients. It astonishes me that such discussion is not only unwelcome in their ranks, but that their understanding of non-invasive holistic dental care is at best – and I say this with great sincerity – pitifully naïve. One dentist shrugged me off, "For what you are suggesting, you might just as well just use water." Not to be difficult, I countered, "Actually, clean water *is* an important healing agent in my natural dental care pro-

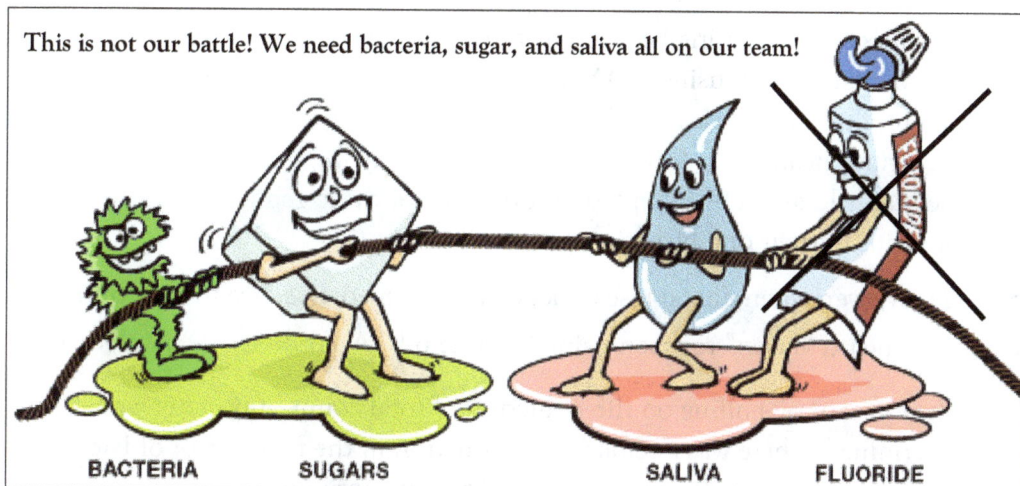

This is not our battle! We need bacteria, sugar, and saliva all on our team!

BACTERIA   SUGARS   SALIVA   FLUORIDE

gram!" Under pressure, they will concede that our saliva is important to remineralization, but with the caveat that it is basically useless without Fluoride-this and Chlorine-that in their entangled war on sugar, bacteria, and the unrelenting but profitable progression of dental disease. Their underlying premise stated boldly in the 3 Tooth Truths makes about as much sense to me as the hypocritical proclamation, "All men are created equal" phrased in the U.S. Declaration of Independence, would have to antebellum African American slaves. Give me a break.

Pulling

I discussed this in the section of brushing, but bring it up again as it is considered a near discipline in and of itself among some dental naturalists. The objective is to saturate the teeth and gums with our healing agents. This is achieved by means of a technique called "pulling."[1] Pulling is simply a matter of vigorously swishing the healing solution all around in the oral cavity, encompassing the teeth, the spaces between the teeth, the gums, tongue and inner cheeks ("buccal mucosa"). Pulling is targeted disdainfully by the dental industry as ineffectual, pseudoscien

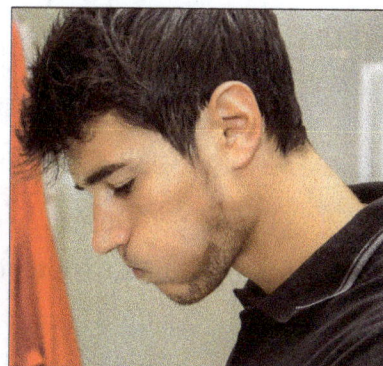

Just swish it around!

tific nonsense. According to the American Dental Association (ADA):

> Based on the lack of currently available evidence, oil pulling is not recommended as a supplementary oral hygiene practice, and certainly not as a replacement for standard, time-tested oral health behaviors and modalities. The ADA recommends that patients follow a standard oral hygiene regimen that includes twice-daily toothbrushing with fluoride

[1]Pulling, also called "oil pulling," is very big among the Paleo foodie factions.

toothpaste and cleaning between teeth once a day with floss or another
interdental cleaner, using ADA-accepted products. Brushing with fluoride toothpaste and cleaning between teeth help prevent cavities and
keep gums healthy . . . If individuals need more help to reduce gingivitis,
they can add an ADA-Accepted mouthrinse shown to reduce plaque and
gingivitis to their oral hygiene regimen.[1]

So they say, even when dental researchers have affirmed evidence to the contrary,
using a standard "Paleo" regimen using "sesame oil." According to Anand, et al.:

> The effect of oil-pulling on the reduction of total count of bacteria was
> determined. There was a remarkable reduction in the total count of bac-
> teria. The process of oil-pulling reduced the susceptibility of a host to
> dental caries. The in-vitro antibacterial activity of sesame oil against den-
> tal caries causing bacteria was determined.[2]

But it doesn't take rocket science to note the ADA's qualifiers, "help prevent" and
"shown to reduce" to realize that their regimens — borne by the statistics laid out
in my introduction — are less than stellar. Not to be drawn into their specious, calculated double-talk, natural dental care is not even interested in their failed
"search and destroy" missions in the first place. Because ADA search and destroy
regimens favor mutan streptococci re-colonization within the tooth structure that
has been stripped of its tartar barrier. Rejecting the efficacy of holistic healing
agents based on the false premise that they don't kill everything in sight is begin-
ner law school debate team sophistry. What they need to concentrate on is that
their regimens simply do not work and come clean about it. But let's not hold our
breath waiting for that to happen.

Our *pulling* objective is to saturate the entire environment with our healing
agents to optimize biological vitality rather than abiotic carnage that favors
aciduric bacteria within the tooth structure. These *agents* help to regulate pH, facilitate healthy blood gases like oxygen essential to natural healing, mineralize the
tartar barrier, and remove excessive mutans streptococci bacteria and their toxic
debris (including collateral damage to other microorganisms) from our mouths via

---

[1] http://www.mouthhealthy.org/en/az-topics/o/oil-pulling (2017)
[2] *Streptococcus mutans* and *Lactobacillus acidophilus* were found to be moderately sensitive to the sesame oil."
Effect of oil-pulling on dental caries causing bacteria. T. Durai Anand, C. Pothiraj, R. M. Gopinath and B.
Kayalvizhi PG Department of Microbiology, V. H. N. S. N. College, Virudhunagar- 626 001, India. African
Journal of Microbiology Research Vol. (2) pp. 063-066, March, 2008.

herbal and mineral transporters. *Brushing* (and finger massaging) aids in stimulating the gums in addition to debriding naturally exfoliating plaque. *Rinsing* with clean water serves to remove depleted healing agents once their tasks are fulfilled while facilitating optimal salivation by hydrating the entire oral cavity instead of desiccating it with conventional rinsing solutions formulated with alcohols. And it goes without saying, in lieu of powerful and destructive Pharma antibiotics, we use a *reasonably natural diet* to strengthen the oral cavity's immune system to mitigate the pathological proliferation of digestive enzymes destroying internal tooth structure, arrest the spread of periodontal infection supporting the teeth, and optimize restorative remineralizing nutrients delivered through salivation.

All of the above may sound like a mouthful of wishful thinking to skeptical readers, and hogwash to the dental industry, but basically we are talking about natural healing mechanisms that were passed on to us in the DNA of our ancient Paleolithic ancestors. The only reason natural healing may sound alien to us, is because the dental industry has conditioned us not to think of it that way, or as a useless anachronism that only applies to long dead cave dwellers. As patients in dental crisis about to lose our teeth to drilling, it is up to us to understand these processes and to put them to work for us, or return to the chair and face the alternative consequences of losing tooth mass.

How much time do I spend pulling?

Pulling, which follows the actual brushing step, is done for 30 seconds to 5 minutes, or perhaps longer if you're a new arrival to my natural dental program and you are experiencing current tooth/gum hypersensitivity or pain. Depending on the severity of the inflammation or the sensitive nature of your gums, add an additional 20 minutes of pulling using raw coconut butter (discussion below). Specific pulling ingredients/instructions I recommend are provided in Chapter 9.

After pulling, the healing mixture can be expectorated or swallowed. Pulling concludes with clean filtered or bottled water if your water source at the sink contains chlorine or fluorine derivatives – typical of municipal water suppliers. These are highly toxic and unnaturally occurring chemicals, and we don't want them in the forms delivered by city water works. Fluoride, which I will take up later in the Epilogue, is very controversial and discouraged by growing numbers of both holistic and mainstream health care advocates as well as more informed communities. Further, its manufacturing has proven to be both harmful and deadly to workers who make it; see Gary O. Pittman interview: https://fluoridealert.org/fan-tv/)

Pulling with coconut butter

The Internet is rife with contentious exchanges over the efficacy of coconut as a useful pulling agent, and, further regarding its safety as a food due to its high fat content. I recommend coconut, period – for food and pulling. Like avocados, I can't think of a healthier natural food to eat and enjoy, and both are a staple and source of healthy fats in my personal diet (Chapter 7). Healthy fats are core to having healthy teeth. How cococont has been relegated to "bad guy" status by some foodists is beyond me, but fingers are pointing directly at the "non-fat" crowd of naysayers who are full of tooth decay like everyone else trying to destroy plaque and tartar. Worth reading is Mark Hyman's, "Coconut Oil – Are You Coco-Nuts to Eat It?"[1] Controversial as a physician in his own profession as I am in my own with healing horses, Hyman details the importance of eating fats and the risks of low fat diets, including heart attacks, cancer, etc. He also debunks the notorious LDL "bad cholesterol" myth and I am in full accord with him. He points out the politics that are involved, but notes that even U.S. government nutritionists are now getting in step and are removing misguided "ceilings" on eating healthy fats.

As a pulling agent, my recommendation is based on relatively recent credible science, and my personal long term experience. According to Peedikayil, et al.:

> The results showed that there is a statistically significant decrease in S. mutans count from coconut oil . . . Coconut oil is as effective as chlorhexidine in the reduction of S. mutans.[2]

Chlorhexidine is the big plaque killer formulated for conventional dental rinses. The good news here is that we are spared the "search and destroy" chemical in our mouths, while the healthful coconut suppresses S. mutans colonization without killing everything in sight.

In the form of coconut "butter," which is what I use and recommend, I find it to be milder on the tissues of my mouth than a daily healing powder/rinse regimen. Thus, it is more adaptable for a longer pulling session for those of you new

---

[1]www.DrHyman.com

[2]"Comparison of antibacterial efficacy of coconut oil and chlorhexidine on *Streptococcus mutans*: An in vivo study." Peedikayil FC, Remy V, John S, Chandru TP, Sreenivasan P, Bijapur GA. *Journal of the International Society of Preventative and Community Dentistry.* 2016 Sep-Oct;6(5):447-452. Epub 2016 Oct 24. Published online by the U.S. National Library of Science, National Institutes of Health.

to my program and who are experiencing tooth hypersensitivity. Many Paleo foodists use coconut oil, which is also perfectly fine, in my opinion, but since we are also introducing a food to our mouths, coconut butter is more nutritious since the whole coconut meat is retained along with the oil. Personally, I find the butter more palatable. Now, you may be asking, how about using unprocessed coconut meat and its "milk" — the "whole" coconut, in other words? Sure! Possibly even better, but then we are dealing with two issues that are solved by the butter: perishability and who wants to go through the trouble of cracking its hard nutshell and extracting the meat and blending in the milk? If you are living in an apartment, the neighbors won't be appreciating all the hammering — and screaming if you mistake your hand for the coconut. I think there's a song here! Whichever form of coconut, find your way to one of them.

Use the organic brand listed in Chapter 9 and follow the use instructions given there. Discontinue once pain subsides and return to the daily use products, which, by the way, are not discontinued; in other words, the coconut butter is used complementary with the regular natural dental care program. Relief should take no more than several weeks. Resume in the future if a tooth "acts up," discontinue once healing brings calmness. Work on diet (Chapter 7) and exercise (Chapter 8)!

As a final caveat, if you are suffering from a massive infection in your gums, then seek out the help of a licensed homeopathic dentist to get it under control before you begin this natural dental care program. Explain to them that you want to use this program, and you might want to show them this book.[1]

Exercise

As I hinted earlier, our natural dental care program isn't for the bed-set depressives. You can't really have a healthy body with healthy teeth if you're just going to sit or lay around all the time "waiting and worrying." That kind of behavior just leads to "pissing and moaning," and the rest of us don't want to hear it. Leisurely walks along the beach, while they may be relaxing, exhilarating or even romantic, isn't going to cut it either. Anymore than bush hogging the back forty on a tractor, because that isn't true exercise either. All that's "leisure" stuff – the en-

---

[1]Only to repeat myself, you must report to your dentist to determine if, to begin, your tooth has been ravished to its inner layers with the liklihood of infection. Natural dental care does not reject conventional dentistry when there has been a long term history of conventional dental abuse. More on this concern in the Epilogue.

emy of strong bodies and great teeth. Even if you have a physical disability, you can do something if there's any muscle mass there and you possess the will to work with it.

Chapter 8 lays out a reasonable routine that I do most every day (I take one day off a week, maybe). It only takes 15 to 30 minutes. At 72, I could still trim the hooves of horses, easier in fact than when I did it in my twenties![1] And it's pretty tough work. Actually it's not, if you exercise and understand the animal. Indeed, after 45-plus years of doing it, it's only possible because I exercise and eat a healthful diet. And I don't whine about the aches and pains that come with age. I do something about it. And that's exercise. There's no excuse and there's no escape. I'll meet you in my exercise room in Chapter 8 after you've finished studying the rest of this book. I can hardly wait to get my hands on your flabby muscles and turn them into something respectable. Be there. I'll be waiting for you.

§

Summary: What to expect from this natural dental care program

It will take several weeks to as many months for you to notice a visible difference in your teeth and gums. Typically, though, one feels better after the first session! But this is just your mouth getting its first welcome taste of a balanced microbiota. Healing occurs at its own rate, and a lot will depend on your own immune system and what steps you've taken to shore it up. Take a look in the mirror now and look at your gums and teeth, but don't get neurotic about it. In fact, stop worrying and go about your life. Your body is healing itself, and this takes time. I waited it out and what a difference! My gums really look and feel healthy. And they and my teeth had really taken a beating by the dental industry over the years. But I've managed to keep most of them, albeit with many fillings and some crowns still anchored to them. There were a lot more coming on and the dentist was convinced I was on my way to very serious problems, as I explained in this book's introduction. But I decided to go this route, trusting my instincts, my research, and, yes, parallel healings I've facilitated with horses in my professional care – documented and witnessed by others.

If you are fighting a cavity or gum disease, consult with your dentist. The pain coincides with the decaying process as it approaches the tooth's root nerve bed.

---

[1]Midway through my 72th year, I retired my trimming business and shifted entirely to teaching new generations of natural hoof care practitioners.

The degraded mass within the cavitation is due to the lactic acid metabolized by the offending bacteria, mutans streptococci, which, in turn, has caused an inflammatory response in the tooth and, depending, possibly your gingiva, too. This is what Cooper was talking about earlier in this book. The healing agents will do several things to get you going on your new path to dental health.

- First, they will begin to regulate the mutans streptococci strains during colonization by helping other bacteria — and your saliva — *prebiotically* mediate the environment of the biofilm so that a balanced community can coexist and flourish together. Not all bacteria will survive to form the plaque platform, but enough of each strain in the right balance will to create a healthy biofilm! *You need them all in balance* for this to happen as nature intends. This stage will partially neutralize and balance the acidification intrusion - very important! Remember from earlier discussions in this book, normal acidification with lesions in the tooth's enamel followed by remineralization of the lesions by your saliva (and our healing agents) is necessary — and perfectly natural — to initiate a thriving biofilm that will transpose to tooth strengthening tartar. This phenomena is a true miracle of nature and what inspired me in Chapter 4 to say the most outlandish thing yet to hit the dental world, "embrace our tooth decay!"

- The healing agents also "clean up" and absorb dead bacteria, cell wastes and other debris. Transporters included in the healing powder rinsing agent (you can also a the probiotic toothpaste, massaging it onto your gums), aided by your saliva, together uptake this toxicity, which, explained earlier, you then expectorate.

- The *powdered remineralizing agents* will also act upon the tooth structure to help your saliva mend surface lesions. You won't be growing a new tooth (yet - Epilogue), but, and the good news is, you may not have to, because new advances in research are demonstrating that the worst cavitations down to the root matrix can be completely remineralized (Epilogue). For this reason, at least the way I see it, not only is this natural dental care important for healing and prevention for newcomers, it buys time for those among us whose teeth have suffered the worst from the industry's practices.

- Last, but of no less importance, rinse your mouth out several times with un-fluoridated and un-chlorinated *clean water.*

Within the tooth, diet will lead the way to mitigate enzymatic degradation and infection by bolstering your body's immune system. I've identified the corridor of healing mechanisms operative deep within the tooth to repair inner structures. The only way to reach them is via the root's vascular system. And this means restoring digestive bacterial colonies in your intestines and delivering "clean blood" to our teeth! As I related earlier, the mere suggestion of this "back door" approach to tooth healing is met with incredulity and borderline belligerence by dentists, who will have no part of it in practice or discussion. Their unequivocal position and practice: root canals. But the supportive research for natural healing mechanisms is there. Our task is to get the diet right, otherwise, the tooth matrix — and any part of our body connected to the vascular "two way highway" — is wide open to Whole Body Inflammatory Disease (WHID) and infection.

All of these agents are identified and discussed in the Chapter 9 with specific use instructions. Be diligent and follow all use-guidelines with the products. Don't forget — brushing and rinsing your teeth with conventional toothpastes and dental rinses described earlier that kill plaque undermine these natural dental care interventions. They are "contraindicated." Use them, and you're on your own.

§

Related concerns

Anyone considering this natural dental care program will undoubtedly have related questions stemming from their current regimen of conventional care. Questions, I have little doubt, they're not going to get meaningful answers from their dentists. Since this is the first edition of *Guard Your Teeth* — and I anticipate future updated editions — I will address those questions I personally raised before creating this program. Like anyone with a lick of common sense, I had answers I needed if I was to free myself completely from the chair and go natural with confidence. When relevant questions arise from readers in the future, I will include them here or, if warranted, in a chapter of their own.

What about fluorides?

Nasty, dangerous stuff. It's the best thing to use if you want to corrupt the microbiotic colonies, which are already in chaos, and muscle up the mutans streptococci up in numbers and strength to keep devouring your teeth. Fluoridated water, toothpastes, and mouth rinses do one thing: wipe out the natural ecology of your oral microbiota essential for creating healthy plaque and tartar fortification. But the "mutans," we've seen, are aggressive survivors and will take advantage of this biological disorder as they repopulate the biofilm faster than their competitors. And you already know what that means from our previous discussions: Lactic acid warfare that no toothbrush, toothpaste, or dental rinse can "reset" to biological order. Proof? Probably the absence of tooth mass in your own mouth. The good news is that you don't need anything fluoridated in your mouth, and evidence for this opinion is strong.[1] Remember, S. *mutans* are simply doing what nature created them to do.

Acid indigestion

Oh, oh! The fact is acid indigestion is a big problem, caused by our lousy diets due largely to our industrialized food chain. If you've got it, then get down right now to your local Walmart or drugstore for some Omeprazole. The stuff is safe and non-prescription. It takes a few days at most to take hold, but once it does, the "pain" from inflammation will go away. Keep using it daily and don't stop unless dietary changes you're going to make render it unnecessary. When you think you're ready, discontinue and see what happens. You're body will let you know when it's safe to go off or stay on this miracle drug.

There are those who will argue that proton-pump inhibitors like Omeprazole are not safe to use, and that they contribute to tooth decay. The evidence isn't there, in my opinion. For persons who produce more than normal stomach acid, and diet doesn't work for them, the alternative to Omeprazole is probably ulcers and maybe stomach cancer. Some people take as many as four tablets a day to control extreme acid production! For them, the alternative may very well be death.

Pharma (Over the counter, prescription, and illicit drugs)

It seems the world today is full of people "on drugs" both legally and illegally. Get off of everything if you can safely, or as much as is possible. You will probably need the help of a doctor and nutritionist if you have a narcotics addiction or a serious disease. Back pain seems to be a growing problem, and it may be that a Class II narcotics (e.g., opioids like oxycodone, fentanyl, and heroin) coupled with an anti-inflammatory agent may be necessary to quell the pain sufficiently so you can sleep and heal — and then get to exercising in my Step 8 gym class. But these

[1]http://fluoridealert.org/

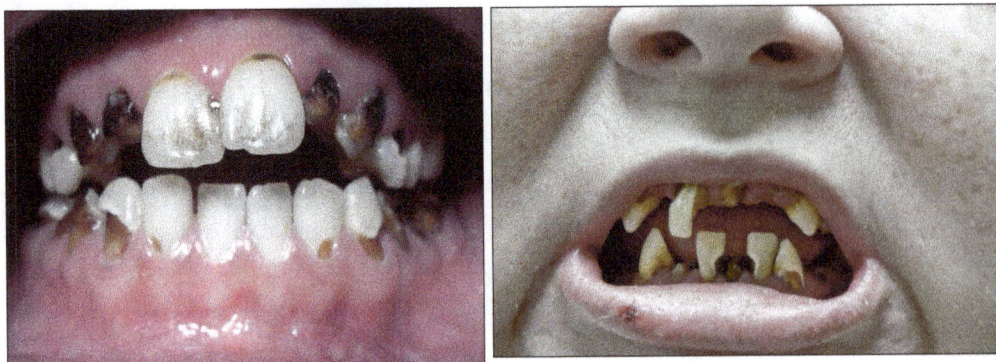

Carious impact of N-methylamphetamine.

class of narcotics can lead to addiction and even death, so work closely with your physician to know when pain due to withdrawal is what you're treating rather than your body's inflammatory response to the cause of the injury in the first place. Obviously, diet and exercise will play an important part of your healing journey because this is nature's way.

The principal concern with Pharma is whether or not the specific drug upsets our microbiota — oral and intestinal — and, therefore, leads to tooth decay (and WHID). To answer the question of "which drug," would require some investigation of existing research, which falls outside the scope of this book.

What to do with your fillings, crowns, posts, and dentures

In spite of what patients are getting on the front lines of mainstream dentistry, in the background, scientists are exploring new frontiers that, if coupled to natural dental care, will revolutionize what happens in that chair. I will take this up further in the Epilogue. For now, my opinion is to try to sustain the hardware in your mouth the best you can. This means diligently following the guidelines published here for natural dental care to prevent further lesions in the oral cavity, and opting for the least invasive procedure possible if any of the hardware in your mouth needs repair or replacement. Remember, the more mass the dentist can get in there to remove, the more at risk your teeth and gums will be for complete removal as the basic tooth structure steadily weakens. Dentists will base the amount of removal on the extent of cavitation, not on the premise of natural healing.[1]

What about my children's teeth?

It is astonishing to me that tooth decay is occurring before the deciduous

---

[1]There is ongoing research suggesting that healing the tooth adds stability to existing hardware adhesion. You'll have to reference the footnote links in the Epilogue and rifle your way through the research abstracts to come to your own conclusion. It seems logical to me, and, so far, is borne out by my own personal experience.

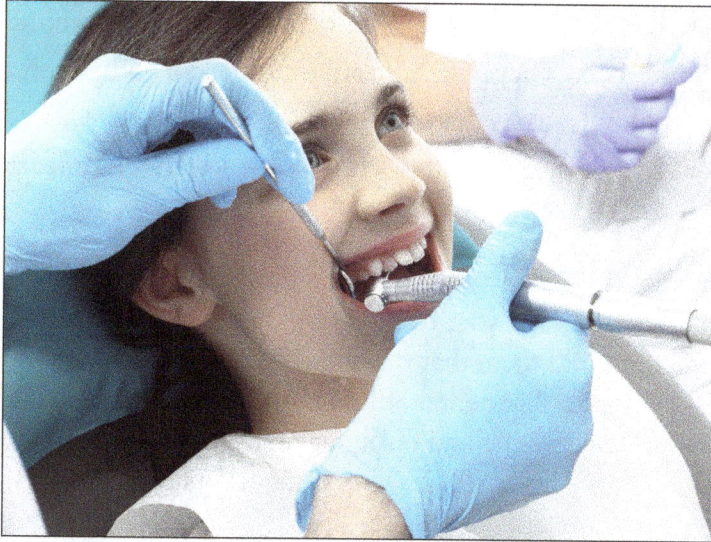

("milk" or "baby") teeth are replaced by permanent teeth. Hence, as soon as lesions occur, dentists will be right in there grinding away regardless of the child's age. It seems no one, from infant to centenarian, is safe from the reach of our industrialized food chain and the dental industry's voracious appetite for tooth mass. The good news, however, is that natural dental care applies to all ages! Teeth are teeth and all are equally subject to the same laws of nature.

My advice is to withhold my recommended healing agents from young children until they are confirmed to pulling and expectorating clean water without swallowing it. Parents will have to make the call. Most children are probably capable of meeting this criteria by the time they begin to shed their deciduous teeth, approximately 6 to 8 years of age. Some youngsters might be confirmable even earlier. At 12-13 years approximately 80% of the permanent teeth are in place, but full dentition is not complete until late in their teen years or early adulthood. Guidelines for children from age 8 are included with the recommended products. There is another product in Chapter 9 I do recommend for younger kids because it is considered a food, and, therefore, is safe to swallow. Although a much preferred alternative to conventional Fluoridated powders/pastes/rinses, it is formulated with the sugar substitute Xylitol, therefore, it's efficacy is somewhat less certain than my recommended adult healing agents which contain no sugars.

What about my dentist?

This is going to be a tough one for many crossovers to natural dental care. My experience with dentists previous to my own crossover are not encouraging. Part of me understands this. It's a profession they believe in and do well with financially. The idea that removing tooth mass is harmful and possibly unnecessary is completely beyond them. They remain steadfast to the 3 Tooth Truths, their drills, and the science and technology that directs their profession. However, a homeopath I know shared an interesting story with me concerning his impact on one dental office:

> A dental office called and asked for me to present to them our mouthwash and tooth powder to their staff, because one of his customers brought it in and told them that this saved their mouth. So I went in and did a presentation to that dental office and they were hooked! They loved it. They now sell/distribute and push our oral care products to their patients (included in Chapter 9). The key thing is, I had to TEACH THEM.[1]

I suspect that natural dental care's acceptance — or intrusion — will occur as a result of patient lobbying, and education will then follow. I'm dubious that dentists themselves will lead the way, although dental related science seems to be several steps ahead of them and going in our direction.

The dentist is only one part of the picture, however. Broadly speaking, the dental industry is very entrenched and financially very stable. I doubt they're going to want to give that up without a fight. On the other hand, I can also see how a shift towards and embracing natural dental care is both possible and profitable. But that discussion takes place in the Epilogue. I'll see you there for that. But first, we still have to tackle the issues of diet and exercising, and here, I expect more resistance from patients than their dentists. But, I'd love to be proved wrong!

Summary

I hope it's clear from preceding pages that natural dental care, while relatively simple to do on a daily basis, and inherently less expensive for healthy daily maintenance when traditional invasive dental care is factored out, will require breaking through the mythology of the 3 Tooth Truths and taking control of our own teeth. While I'm not holding my breath that current generations of dentists will see the light and embrace this new path to dental health and help, it's clear to me that some will, and, in fact, some are right now. They promise to be the new generation of professional natural dental care practitioners. I'll address this further in the Epilogue. For now, we need to address more information that is needed if we are to go it alone in the meantime.

[1]CAPS are his emphasis!

Chapter Seven
"Feeding" Our Teeth

A reasonably Natural Diet! Diet is a big one, because if we don't get it within the realm of "reasonably natural" there is no way to stop the aciduric mutans streptococci's destructive path through the tartar barrier. We must feed them what they want and need, or they will drill into us with such zeal as to give any drill happy dentist an inferiority complex. For the mutans streptococci, and maybe other bacterial strains as well, *this means sugar.* And either we give it up to them, or it's war. And, so far, witness our teeth, they know all about winning dental wars. But, and I hate to have to bring this up again and again, these sugar eaters aren't the only troublemakers inhabiting our bodies. That's right, we're back to our equally out of control sugar hungry intestinal bacteria that cause Whole Body Inflammatory Disease (WHID), partners with mutans streptococci in creating the "two way highway" of destruction. The only way I'm aware of to "negotiate" a peaceful settlement with them, is through diet.

We've seen nature's defenses deep within the tooth structure where the microbial invasion has penetrated and it's up to us to come to our tooth's rescue via *diet* — our only way in to make peace and gets things back in good order. The dentists can't help us, because to them, diet is a vague "Tooth Truth," mythical nonsense that our aciduric bacteria love to feast on! Recalling the wise words of the Sioux healer, Little Crow, this is a mess we ourselves have to figure out and clean up because "no one's coming down from anywhere" to do it for us. This isn't Indian jive, this means drawing a red line in the dirt that all these aggressive bacteria don't get to cross. They get just what they need on one side, we get what we need on the other.

The Paleolithic diet conundrum

I've already said I don't want to get squeezed between the embattled food extremists of the modern Paleo and Veggie factions. We'll cut our own path through both of them. We have to. Because our Paleo ancestors were omnivores — they instinctively ate anything derived from animal or plant life, and minerals (substances neither animal nor vegetable) that delivered the nutrients and energy they required to survive with vitality. Our species today has neither mutated nor evolved away from the dietary requirements of our inherited *H. sapiens sapiens* Paleolithic

DNA. It's what it is, and we have to live up to it.

Insofar as the specifics of the ancient Paleolithic diet goes, I've learned that much of mainstream dental science just has it wrong. But the truth is, so do most people, in and out of science. So, I'm not surprised at all that dentists and their brainwashed patients are both as pitifully informed about the truth of the matter as they are of the tooth's pathology-free natural state. The meager "meat, nuts, and berries" dietary stereotype of Paleolithic *H. sapiens sapiens* is simply unadulterated nonsense. It's the parodistic dinosaur baloney of vintage 1960s Flintstones cartoons. Apparently people still believe that was a reliable archeological representation of our ancient ancestors. We've got to flush all that BS out of our minds along with the dental industry's "get back in the chair" threats.

But I would also go so far as to say that the Neolithic agricultural revolution isn't really as bad as some might think. Cooper himself didn't say it was bad, only that we have to learn how to feed our microbiota in spite of it, or perhaps in conjunction with it is better stated. What has distilled down into the 3 Tooth Truth mantra, however, is that the modern agricultural revolution brought us refined sugar, and that we are addicted to it, and that's end of the miserable tooth story. But we know that's not exactly true. We know that sugar and tooth decay are essential for building strong teeth. How I'd like to tuck that mantra under the rug and out of sight myself! But I can't, even though to critics of natural dental care it puts me and my credibility on the 3rd Floor, 1st Ward, second bed on the left, in the Paleo "nut house." *Sigh.*

Paleo couple Fred and Wilma Flintstone weren't beneath advertising cigarettes during the show's closing credits, breathing more lifestyle nonsense into "caveman" mythology and the uncritical minds of millions of viewers being brainwashed into believing that smoking cigarettes was a Paleo habit worth keeping. But our decaying teeth, lung cancer, and emphysema say otherwise!

What I've come to realize is that trying to unravel the giant ball of speculation and misinformation patch-worked together in the 3 Tooth Truth mantra is just a waste of time. But for the sake of people looking for actual solutions to the cavity debacle, and using natural dental care as their platform, I think it's important to sift and sort through this business of the Paleolithic diet. This means evaluating the blessings and problems in our industrialized food chain, and coming to an un-

derstanding of our natural dental care's concept of a *reasonably natural diet.*

My interpretation of the ancient omnivorous diet of Paleolithic *H. sapiens sapiens* is based on the archeological evidence that it delivered sufficient vital nutrients for our specie's vitality and survival. If that weren't the case, I wouldn't be here today to write this book, and you wouldn't be here to read it. But natural selection does not entirely stand still. We are hundreds of thousands of years since the dawn of our species, and clearly there are specialized adaptations that have occurred along the way to coincide with our progression from strong-jawed hunter-gatherers to gracile-jawed sendentarians. Indeed, the size and shape of our jaw bones is one example of this adaptation, which I'll discuss further in Chapter 8. Nonetheless, we are still the same species. And it should be clear to all that our species was intended to coevolve technologically and intellectually. Naturally, our diet was destined to change because of this. And it has. But good science would argue that both the Neolithic and modern agricultural revolutions were important to our survival as a species. Indeed, our innate propensity to innovate such things as food production may very well have prevented our own extinction from famine and disease. Nothing to take lightly!

But this is not to say that our early Paleolithic diet was tasteless or bitter, dull and boring. And so deficient as to render our bodies malnourished. And that people died left and right from eating the wrong things because it wasn't obvious to them what was toxic and what was safe. People today actually believe all of this. And I think it's a combination of lousy science, misguided public education, and Hollywood nonsense that has taken us there. Didn't we all believe that the Flintstones enjoyed barbequed dinosaur steaks, even though the last incarnation of dinosaurs became extinct 60 million years before humans arrived on the planet? Give me a break! In spite of all the Hollywood smoked dinosaur sausage, and in spite of what naysayers and bad science have taught us in our sleepy, regimented public schools, our Paleo ancestors ate well and, gauging from their great teeth and robust bodies, likely suffered with less malnutrition and tooth decay.[1] Some point to "wild foods" as the reason, pointing, as an example, to extant, luscious, and sweet wild fruits that were and are palatable to our species.[2] While early *H. sapien sapien* migrations no doubt coincided with food shortages, along with regional family band or

---

[1]I've already shared data on tooth decay today. According to estimates by the Food and Agriculture Organization there were 820 million under- or malnourished people in the world in 2010.
[2]The Flintstoners would have been delighted. For a small sampling of wild fruit delicacies and available, go to —
https://deniseminger.com/2011/05/31/wild-and-ancient-fruit/

tribal conflicts, curiosity and the desire to explore the unknown — and to innovate — were, I'm certain, the primary motivations. They were human beings destined to become us — not some alien species out on an ephemeral fling on the planet earth! We don't stand still, the mantra of our DNA is to progress!

Of long standing interest to me that has relevance to the adaptive progression of our specie's behavioral ecology that definitely impacted diet, is another battleground of dissent among scientists: "Anatomically modern humans (AMH)" versus "behaviorally modern humans (BMH)." In keeping with our specie's earliest incarnation 300,000 or more years ago, if we were physically built the same way as we are today (i.e., AMH), then did we not have the innate *cognitive ability* to behave the same way as we do today too (BMH)? One side says yes, speculating that the modern behaviors of today resulted from an accumulation of cultural experiences, not a limiting gene awaiting evolution in our DNA's double helix. Whereas the other side contraposes that a genetic shift in cognitive abilities occurred much more recently (50,000 years ago), supported by archeological evidence (e.g., the sudden and widespread appearance of tools). If the latter, are we then to believe that our "lesser developed" minds with perfect teeth, suddenly gave way to "more developed" enlightened minds that paved the way for mouths full of tooth decay, nuclear bombs, and global warming? I don't think so. That's like saying, it took 250,000 years before our species could develop a language to communicate with. Hard to believe. But even there we have a hot bed of debate among scientists! I suspect that there was a whole lot of "talk" going on among our Paleo ancestors from the very beginning! But what form that talk took no doubt varied as widely as their experiences in life and where they lived. The rich plethora of extant languages today "speaks" to this probability.

As the modern industrial and agricultural revolutions consumed and replaced the Neolithic and lingering Paleolithic cultures, not a few surviving tribal peoples still successfully hunted, gathered and domesticated wild plants in remote, pristine areas. And still enjoyed great teeth! I've cited examples in Chapter 1 and elsewhere. In the newly "discovered" Americas, European colonists adopted, and even expanded on native species, creating new varieties we enjoy today. But then, as Cooper discovered, things began to go wrong. Terribly wrong, and it showed up in our rotting teeth. Was it that the coevolving modern industrial and agricultural revolutions were inherently "bad?" I don't think so, at least within the context of an internationally expanding civilization bent on progressive development. And history clearly shows that other factors entered the picture that would explain emerging dietary problems occurring broadly across humanity.

For example, as societies became more and more agrarian, giving rise to slavery and feudalism, diets clearly deteriorated for the poor. Displaced native peoples' lives and diets also corrupted. And as feudalism gave way to cities, more and more people were removed from the land. Socialism and communism soon clashed violently with capitalism over what was the best way to feed the burgeoning impoverished and landless masses. At the same time, science and industry increasingly mechanized farming, impacting the lives of animals, plants, and people. Corporate farms soon began replacing small farmers, many of whom lost everything, including their connections to the land. New commercial feeds intentionally or inadvertently favoring WHID in animals were formulated. Powerful chemical fertilizers, dangerous herbicides (such as Agent Orange) and pesticides were also used everywhere to sustain production. It was just a matter of time before these chemical and biological agents also found their way into our bodies and our mouths. Metabolic "diseases" like cancer erupted ubiquitously both among farm workers and anyone who processed or ate contaminated food, meaning, "the rest of us." Farm animals also met the same fate.

While the foregoing was evolving, manufacturers unlocked sugar from the plant, creating "free sugars" for a multitude of culinary applications. In time, they also created artificial substitutes in the laboratory. Because we're all easily addicted to sugar, industrialists — now basically controlling our food chain — began putting it in just about everything we ate, drank, and otherwise put into our mouths, including our toothpastes! Good for all the aciduric bacteria in our bodies, but bad for us.

None of the above, of course, is "good news" in the world today among those even remotely concerned with the nutritional value and safety of things they are eating and brushing their teeth with. The rise of the natural foods movement, organic growers, and alternative (holistic) medicine were born of, and in reaction to, this concern, as was this natural dental care program. It now seems logical, and doable, to sift out the toxic things from our diets (and lives) that favor microbiotic disorder and proliferation of harmful strains to the detriment of others. All we need from our food are those nutrients that our ancient ancestors needed for their vitality. In principle, I would go so far to say, eradication of oral disease should simply be a matter of doing just that — consuming a reasonably natural diet that serves our specie's vitality, and nothing more. But because of human innovations in science and technology, along with the good, we have to make smart choices to eliminate or minimize the bad. I know, easier said than done, given our industrialized food chain and all its marketing hype!

For now, here's what I do and what I recommend for a "reasonably natural diet." One that should satisfy each of us ("I") and our digestive bacteria ("the rest of us") as we navigate our way together through the maze of our institutionalized and industrialized food chain.

Go organic!

Go "organic" when you can, if not, then non-GMO, and if that's not possible and you can't grow your own, then go as "fresh" as you can get it. The less processed the food is, including those grown with pesticides and herbicides if you have no other choice (hard to believe today), the better off you will be than eating out of a can, TV dinners and other prepared meals, and fast food fares, all laced with more chemicals than actual food ingredients. But even here, we now find organic versions lined right up next to the other on supermarket shelves, restaurants, and even the fast fooderies.

In recommending the organics food pathway, I would be remiss in not saying there is considerable contention between those who proclaim "organic is best," and their adversarial doubters, who say, "Phooey!" So, a brief diversion:

I've been a lifelong admirer of the late great health and body building buff, Jack LaLanne (more on him in the next chapter) who enjoyed debunking the natural foods factions as being dead wrong on organically grown food. The only difference nutritionally between the two, he claimed, is that organic costs more. LaLanne was also a semi-vegetarian, I believe a *pescatarian* factionist.[1] On the other hand, our "expert" ancient Paleo ancestors with great teeth were organic all the way, but they were also omnivores. How many of them aged to 96 like Jack, who vigorously exercised until the day before he died, is still anyone's guess. My Uncle George, in contrast, who seldom exercised and occasionally used organic, lived to 101 with excellent health – but he was a great chef, using an extensive variety of fresh foods. I've pretty much modeled what I eat and how I cook after his example mainly because his cooking was so damned good, although, as I say above, I think going organic when you can is the better, safer path to take, particularly if you are suffering from life-

---

[1]"Pescetarian" or "pescatarian" is a neologism formed as a portmanteau of the Italian word pesce ("fish") and the English word "vegetarian". Both fish and shellfish are included with their vegetarian fare.

threatening forms of WHID, such as an aggressive cancer. Both men were born the same year, 1914, but George outlived Jack by five years. Another coincidence: both died of pneumonia. Although in many respects Jack has been an enduring role model for me as he was an activist for what is "natural and healthy," so was my uncle in his own way, and I find myself pointing equally to both of them with gratitude for their contributions to healthy living and personal inspiration. My horse-related books are all about vitality too, for which I've been characterized as a "nut" and "trouble maker" in many courts of opinion across the horse-using community. How delighted I was to read this comment though from Jack, "People thought I was a charlatan and a nut. The doctors were against me — they said that working out with weights would give people heart attacks and they would lose their sex drive.[1]" *Ho-hum.*

Free sugars

Overall, you want to moderate "free (i.e., table) sugar" use, but not necessarily eliminate them entirely. Of course, telling anyone in the world today that you have to give up sugar to have healthy teeth would insure this natural dental care program of an early grave. They'll take the chair first. This is because any sugar, free or in fruits, will feed *S. mutans* to its greedy satisfaction, we just need to use some calculated discretion in how we deliver it. On the other hand, I'm less certain that artificial sugars are a safe choice, for the reasons I explained in Chapter 6, and so I will discourage their use here. You're on your own if you go that route.

I use organic honey, organic cane sugar, sorghum, and organic bootstrap molasses. I classify all of them as "table sugar." I use them all moderately though, tablespoons at most in a day, if that, although at times, even too many times now and then, I really indulge. Fructose is also offered as a table sugar, and is relatively safe (although not all scientists think so) — unless you are diagnosed as *fructose intolerant*, then don't use it at all. But all of these "table sugars" contain sucrose and are digested by *S. mutans* into lactic acid during the formation of dental plaque, or biofilm. We recall from Chapter 3 (p. 24) that nature through salivation firsts lays down a film that facilitates biofilm adhesion and formation.[2]

Complicating the sugar picture further, unfortunately, is that sugars are added into just about every processed food item in the grocery store. There's so much in

---

[1]Goldstein, Richard (24 January 2011). "Jack LaLanne, Father of Fitness Movement, Dies at 96". The New York Times. 24 January 2011.

[2]Called *dental pellicle*, a protein film created during salivation that forms on the surface of enamel, dentin, artificial crowns, and bridges. It is the first layer of protection from the acids produced by oral microorganisms after consuming carbohydrates. Biofilm forms next, which is then calcified by saliva into tartar.

there, I can feel ol' Mutans peering out of my mouth and licking his slimy chops as I'm walking down the aisles. Even the "natural foods" store has become a "Mutan Heaven" these days. And you'll find it also in all the fast food and upscale eateries, coffee bars, and lemonade stands mommy helped us set up as children to make extra change and get us out of her hair. Everyone, and I mean every *body*, "Mutans and Humans" alike, all love sugar. No, this isn't love. Let's call it what it is: an *addiction*. Poor ol' Mutans is as addicted as us. From birth, we and our fellow little Mutans babies are raised together eating it. So, we're all in this together. But whatever your source, be sure to use sugar in some form, always organic, and always in moderation, to aid S. *Mutans* lay the necessary foundation for healthy plaque formation. That's their job, and they're not about to let us put them in the unemployment lines.

There is a current, although minor, movement to label all free sugars as a non-food and a dangerous drug that causes addiction, disease, and death. At face, this is hard to argue with because sugars, all sugars natural or otherwise, do feed mutans streptococci that can lead to tooth decay, and are implicated in diabetes and other metabolic disorders. Not a very good reputation, and harder to defend! But sugar in some form is necessary for bacteria to create healthy plaque and tooth fortifying tartar. Even fresh fruits (discussed below) are metabolized by these bacteria into sucrose (table sugar). In its defense, sugar is also a source of energy, and a way to enhance the palatability of many foods that are nutritious. Depriving our bodies of sugar can lead to serious, even life-threatening complications associated with hypoglycemia ("low blood sugar"). It is clear, nature intends for us to eat sugar in some form. So, we might as well enjoy it! I know I do, and I still have my teeth and intend to keep them!

Given this outlook, which admittedly makes me look pretty bad as the author of a purported natural dental care program, I have no choice but to counter the indictment of free sugars as being misguided, or at least misguiding in its myopic view based entirely on blatant sugar misuse that truly leads to metabolic pathology. As an advocate for the safe use of sugar, I will make my case and recommendations in the Epilogue.

Fresh fruits

Fruits — along with honey, tree and vine fruits, flowers, berries, and many root vegetables — bear the sugar fructose. But S. *mutans*, which we've seen requires sucrose to build plaque, is able to cleave (metabolize) fructose molecularly to access its sucrose, and we all know what that means now. On the bright side, while both fruits and table sugars give us energy for work, fruits deliver many other vital nutrients nature intends us to get from them. For this reason, it can be argued that they are a preferred choice to the free sugars to feed our bacteria. Personally, fruits play a principal and vital role in facilitating healthy plaque as part of my daily morning smoothies. As a nightcap, I have been known to down fresh-squeezed organic orange juice with popcorn almost every night, along with several tablespoons of organic whole milk yogurt on the side to shore up my probiotic environment. I can't help myself. I'm a food addict!

Vegetables, legumes, cereals, and grains

Here, there are warnings raised by both Paleo and Veggie factions concerning potatoes, rice, and beans because of their phytic acid content, which is alleged to transport vital minerals like calcium, iron, and zinc from the body due to the acid's molecular binding affinity. In my opinion and from experience, other nutritional benefits far outweigh and override any such concerns. And if you eat a balanced diet of many natural foods, there absolutely isn't any concern. I try to eat as many vegetables raw as I do cooked (usually steamed or slow cooked in my rice cooker).

Exercise restraint if you can with commercial cereals as they're pretty much loaded with those free sugars these days. I'm particularly fond of oatmeal with cream, butter, fresh fruit, nuts, and honey or bootstrap molasses.

Fats and oils

Use "real" butter for cooking, as a spread, and as a condiment ~ instead of the margarines. Eat animal fats freely as they naturally come in the flesh. Use chilled vegetable oils (e.g., in salad dressings) and minimize using them heated, substituting butter or animal fats instead. These are all healthful and natural for humans. In fact, I'm cautiously dubious that we can enjoy optimal health and freedom from tooth decay without them. Regulate quantities to how active your lifestyle is – just make sure you eat them and don't buy into the "non-fat" and "low fat" versions that are now rampant in the grocery stores. They have become so ubiquitous, I really have to hunt down the "whole milk" varieties. The low/non fat foodists will fight you to death on this. Ignore, ignore, ignore!

Eggs

When I feel like eating them, which is usually several times a week, I eat as many as I feel like, usually two. Use butter or water to cook them, or don't cook them at all. I also use them raw (if organic) now and then in smoothies, usually when I make my organic version of the popular "Orange Julius" with fresh-squeezed oranges and fresh cream and vanilla.

Dairy

As far as I'm concerned drink and eat all you want or feel like having. I love all the dairy products, and like good coffee, I can't imagine life without them. Use the "whole milk" versions to get their healthy fats, and avoid low and nonfat forms, which don't even begin to taste as good anyway although people try to convince themselves of that. Regulate the quantities of whole milk you drink based on how active your lifestyle is. The less active, use a higher ratio of clean water as a substitute; the more active, use proportionately less water. But when you enter my exercise camp in the next chapter, "fat" concerns will melt away.

I use whole milk yogurt and kefir — and sourdough bread — as my principal "probiotic" allies to help counter and balance the S. *mutan* strains in my mouth and other bacteria in my intestines. I drink quart of organic whole milk each day, along with salted butter according to taste, and a variety of animals fats as part of my flesh eating regimen. As I've gotten older, and not quite as active as in my earlier years, I've "down-sized" my milk allowance gradually from 2 gallons/day 15 years ago. According to "experts," I should have been dead years and years ago. But, as you'll read more about me in the exercise chapter, I'm anything but a "dead beat."

Animal flesh and organs

The vegetarians and vegans may wish to bypass this section, or grab hold of some Omeprazole to combat a bout of acid indigestion. It's bad enough that I had to send my vegan dairy critics to the trauma ward. I eat any flesh food that my body in the moment "tells" me to eat, which usually means one or more servings every day – usually beef, poultry, fish, and pork. When I feel that "crave," I go for it. Indians and pioneers hunted in the woods and at sea. I hunt in the grocery store and take the hunt there very seriously. I don't eat wild because I don't like the taste of it, except for fish. My backwoods clan evolved out of the holler long

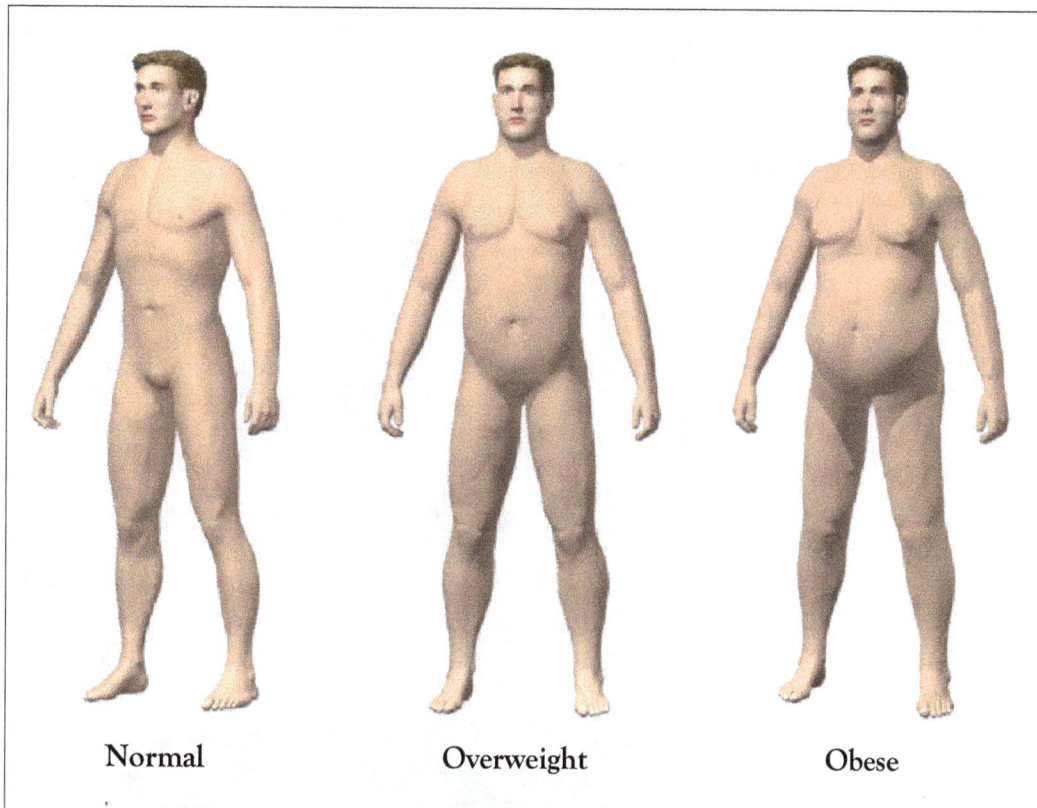

| Normal | Overweight | Obese |

ago, and we can't stand the taste of wild boar or whatever's roaming around out there; in fact, like the gentle vegans and veggies, I don't believe in killing wild animals for sport or any reason. That is, unless you're living like a Paleo and that's all you've got to eat out there off the grid in the Arctic Outback, on some mountain, or down in the holler. I've got no problem with that.

I cook flesh foods over fire, in the frying pan, or in the oven. Whichever way, it's got to be cooked. I don't eat raw flesh food, and can't stand sushi. I scout out flesh that is well-marbled in fat and preferably has the bone with marrow still attached to it. Eat less marbled flesh foods, however, if you are less active. I'm very big on ribs, juicy cheeseburgers topped with cheese and fresh vegetables, simple stews, most any steak, baked chicken, baked turkey, meaty lasagna with spinach noodles, and most seafood. My chop suey ranks among the best according to friends. I'm getting hungry as I'm writing this. The truth is, like my late Uncle George, I love to cook and I take it pretty seriously because I love to eat great tasting, nutritious food. Now I'm really starving. I don't limit the amount I eat in a sitting except to stop when I feel full and satisfied. If you exercise as I do, you will probably eat less because your body metabolizes things differently than if you are a couch potato laying on the sofa watching TV, growing fatter and more sluggish

Devastation of Whole Body Inflammatory Disease: (*Above*) Laminitic bovine hoof; (*Below*) laminitic equine hoof. *White arrows* point to cascades of stress rings growing down with the hoof wall. These are symptoms of long term or chronic WHID. *Blue arrows* point to catastrophic splitting of hoof walls. Black arrows point to pathological separation of entire hoof wall from foot.

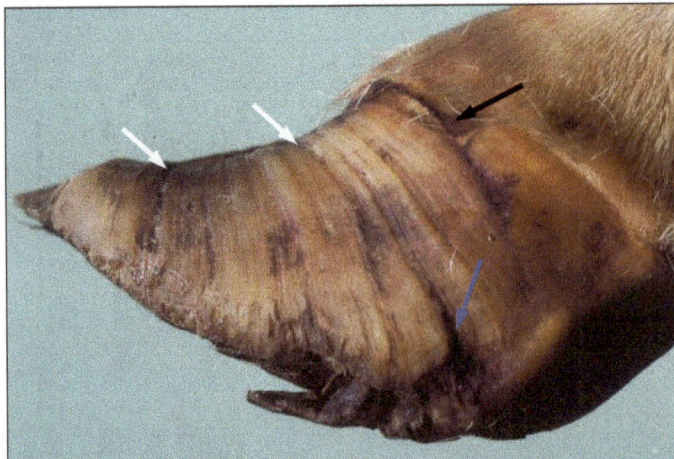

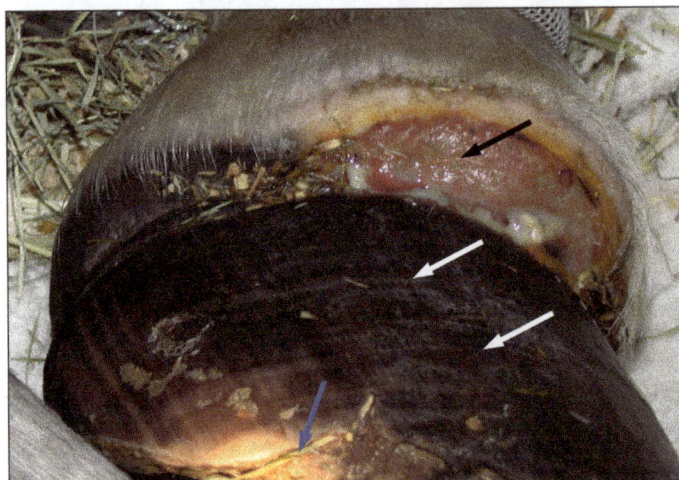

with every meal, aching with joint pain, pissing and moaning. Your mixed-up body then demands that you eat more and more and more to quell the incessant pangs of hunger to fill your gross, ever-expanding stomach.

* * * * Food "Red Meat Alert!" * * * *

I do want to caution against what many today are considering to be a "more natural" and healthier way to raise domesticated animals for consumption, including bovine, swine, and other edible ungulates (hoofed animals, and in many countries that includes horses!). This is what is touted in both supermarkets and natural foods stores as "grass fed" livestock. These are animals that are raised exclusively on ("grass") pastures. This is pretty much the direction of the natural foods industry, although its now definitely creeping into the mainstream supermarkets too. I understand, and am sympathetic with this direction, because, in large part, it's a humanitarian response to the mainstream's "finishing plants," where livestock is practically "force fed" with sugar-rich grains in crowded, traditional feedlots that stink to high heaven for miles. Nonetheless, my opinion is that both practices — grass fed or feedlot — are inhumane, but for different, although related, reasons. Let me explain, because I believe there is a solution — and a good one!

Like horses, cattle that are pasture raised or "finished" in feed lots develop what is called *laminitis* of the hooves (*image above*). This is a very serious problem that can, and often does, lead to incredible suffering and death of the animal. Technically, laminitis is defined as a painful inflammation of the sensitive structures responsible for connecting the hoof to the hoofed animal (horse, cow, swine, etc.). Early symptoms vary from *sub-clinical* (no pain) to *clinical* (acute pain) to *chronic* (long term fluctuations between chronic and clinical symptoms) that are often attributed to unrelated causes (e.g., changes in the weather, arthritis, and allergies) rather than unnatural diets. Sugar — typically molasses, beet pulp, and cane — are the principal causal agents ("triggers") for laminitis. But laminitis is actually a symptom of Whole Body Inflammatory Disease (WHID) — that's right, the same disease that affects us and our teeth. The difference between our species being — or so I thought — the variant strains of pernicious bacteria. In horses, cited in Chapter 1, two of the "sugar hungry" intestinal bacteria are *Streptococcus bovis* and *Streptococcus equinus*.[1] But, *S. bovis* is also showing up in humans,[2] where it is also classified as a pathogenic germ, often associated with colon cancer[3] and other catastrophic manifestations of Whole Body Inflammatory Disease (WHID). In my book *Laminitis: An Equine Plague*, I connected the dots between ourselves, the animals we eat, and the likelihood of cross-species contamination via these germ vectors:

> Given that many of these species are part of our own food chain, are we not wise to ponder the implications of laminitis in relation to our own health? In fact, these are sick animals that are infected, contaminated, and hosting strains of bacteria — and very likely predator viruses suited to their environment — that belong in no living creature.[4]

It's not hard to imagine, then, that livestock diets rich in fermentable carbohy-

---

[1] *p. 6, fn 1. Equine Laminitis.* C. Pollitt.

[2] Corredoira-Sanchez J et al. (2012). "Association between Bacteremia Due to *Streptococcus gallolyticus* subsp. gallolyticus (*Streptococcus bovis I*) and Colorectal Neoplasia: ACase-Control Study" (PDF). Clin Infect Dis. 55: 491–496.

[3] Klein RS, Recco RA, Catalano MT, Edberg SC, Casey JI, Steigbigel NH (13 October 1977). "Association of *Streptococcus bovis* with carcinoma of the colon". N. Engl. J. Med. 297 (15): 800–2.

[4] *Laminitis: An Equine Plague.* J. Jackson. NHC Press. P. 10

drates are culturing pathogenic intestinal germs that are transferring to our own bodies. There appears to be evidence for this, and that this contamination is sufficient to propagate microbiota disorder anywhere in our bodies, including our teeth. To put it bluntly, we are feeding our livestock to rot our teeth. Not a good thing!

Can ranchers and feedlot operators do something about this? Absolutely! First, they need to recognize the sub-clinical symptoms of laminitis. In the photos on the facing page, the arrows point to what NHC science identifies as "stress rings." These are visual indicators that the animals are infected with bacteria that thrive on sugar. Second, livestock should be removed from grass pastures as these are invariably rich in sugars due to photosynthesis, and are triggers for laminitis and WHID. Third, feedlots and other close-confinement feeding systems should be shut down, not just for their foul-smelling inhumanity, but because feeding strategies are aimed towards "fattening" the animal for slaughter, and this translates once again to more sugar, and, therefore, laminitis and WHID.

So, with the problems now identified, what is the solution?

My book, *Paddock Paradise: A Guide to Natural Horse Boarding*, provides a genuine solution based on equine life in the wild that works for just about any domesticated species, particular those with hooves.[1] Specifically, it lays out management guidelines that facilitate natural feeding behaviors seen among wild horses, controls on safe feeds, freedom of movement 24/7, and natural socialization. All of these are important to the vitality of any species.[2] But I also wrote the book in response to horses living in unnatural confinement systems, and also suffering epidemic levels of laminitis and WHID. These are the same problems befalling so many large animal species living in human captivity. Paddock Paradise is based on sound ecological principles that do not harm the environment either.

Desserts!

I have always been a great dessert eater, like most people I can assume. They are high caloric and full of sugar as a rule, so if you are going to eat them best

---

[1] Available online at Amazon, Walmart, and other book outlets. See also: www.PaddockParadise.net

[2] I've also targeted wild animals in captivity in my related book — *Zoo Paradise: A New Model for Humane Zoological Gardens (2019).*

that you've got your weight down where it belongs through exercise, vital signs in check, and your teeth aren't a disaster. My other piece of advise is to make them yourself so you know what the ingredients are. And then moderate your servings. If you can't do that, then move to fresh fruits for your sweet tooth!

Clean drinking and cooking water

Very important that it be clean – meaning filtered and free of Chlorine and Fluorine derivatives. If your water system isn't filtered at home, then use bottled water. I drink clean water when I'm thirsty, and preferably chilled when I drink it with my meals, working out, or working with horses. I also use clean water for cooking and, of course, as a core part of our natural dental care regime. How it's used in the program is discussed in Chapter 9.

Soft drinks

Like millions of people, I love CocaCola and other soft drinks. But knowing now how conducive they are to dental caries, be frugal about it. I limit it to only one now and then. The same goes for sweetened tea, hot or iced. Otherwise, try sparkling water. When eating out, most of the time, I order milk – usually I'm the only person in the restaurant doing so these days, when alcohol and soft drinks (and increasingly bottled water) seem to be the natural order of the day. For whole fruit juices, I use my juicer (retaining the pulp to eat for the fiber); or I eat fruits whole. Raw fruits also go into my morning smoothie. Speaking of which . . .

Smoothies

I prepare a morning smoothie when the urge strikes. My foundational ingredients include bananas (almost daily), papayas, all sorts of melons (including watermelons), berries, dates, raisins, and nuts. I vary all of these by day, depending on which ones I feel like eating and are seasonally available. Into this base goes a half cup of sugarless whole milk yogurt or kefir, and a half cup of clean ice. Grind the whole thing up and take it down. I'm a confirmed, unrepentant, unrehabilitatable smoothie addict. All my bacteria love it in what can only be a "grand harmonic convergence." It goes without saying, go with organic if you can get it.

Coffee

I would rather die than not have great coffee. Good coffee is nutritious and important in my natural dental program. I always drink it black. And now, a story:

I've been a Peet's Coffee aficionado and advocate since Alfred Peet opened his first shop at the foot of U.C. Berkeley in 1966. Ironically, I was introduced to the store by a college mate who went on to become a dentist! I've proudly created many Peetnik addicts over the years. We would all cry, scream, gnash our teeth, and threaten suicide if we were cut-off permanently from our Peet's coffee. Going without Peet's even for a day makes us neurotic and guilt-ridden. My morning rituals always begin with Peet's, no matter where I go in the world, and I always take a supply with me to feed my addiction. Starbucks got its start with Alfred Peet's help, and its founders (trained by Alfred) and top execs hold Peet's up as their great inspiration.[1] In fact, the entire current gourmet coffee craze in the U.S. can be traced to Alfred. Well, sort of. My uncle George, an artist living in San Francisco after WWII often took me into the beatnik coffee houses along the Embarcadero during the mid-1950s, where other artists, poets and freethinkers met to gossip, share their art in whatever form, and imbibe the exotic coffees and Turkish blends that the worldly shop keepers found and served to their eclectic crowds. Alfred Peet was there in San Francisco during this period, and while he learned his trade in Europe before the war, he realized like my uncle that America didn't really have easy or broad access to good coffee because it didn't exist in the new post-war generation of supermarkets. He crossed the SF Bay to Berkeley and did something about it, and the rest is history. Years later, George and I were chatting over some good coffee when I brought up Peet's story and his coffee. "No big deal," he retorted, "we all drank great coffee like that back then." Right, of course!

Things better not to put in our bodies!

Alcohol

There's no good reason to drink the stuff, although Jack LaLanne loved his red wine as did my uncle. Jack was actually ticketed for a DUI in his home town

---

[1]The only significant blemish in Alfred Peet's life occurred during WWII, during which, as a young man, he donned the uniform of the German Wermacht (Peet was Dutch, not German) – in effect, collaborating with the Nazis while his fellow countrymen were brutalized and starved by a German military occupation. His disappointing and arguably disingenuous rationalization, and I'll leave it at that, was "I was inspired by their 'goose steps' during parades." During the same time, my Uncle George at age 30 was an Army sergeant whose unit was fighting the Germans near the Ardennes Forest, otherwise known as the "Battle of the Bulge."

of Morro Bay a few years before he passed away. *Sigh.* If you do, be stingy enough with it that you can't remember the last time you had a drink. Alcohol is a great killer that has ruined countless lives and created as many family tragedies at the hands of drunk drivers. So, be careful. You know you're an alcoholic or on the way to becoming one *if* "you have to have a drink." I personally don't care for alcohol, with the exception of a traditional blended Margarita if we're out to eat somewhere that serves it. My imbibing average is about 1 Margarita every 2 to 3 years, although I can't remember my last one and there's no urge to go get one.

Tobacco

Careful, too, the stuff is dangerous, and evidence is incontrovertible that it can only harm our bodies, including our teeth and gums. I do not smoke and do not recommend it.

Drugs (legal or otherwise)

First off, I'm not partial to legalizing nor making illegal marijuana or any drug used medically, or for divertissement, believing that people should make up their own minds. But I am in favor of decriminalizing drug addiction (whatever the drug). This is a tragic medical problem, in my opinion, and, other than for the purpose of illegal narcotics interdiction, law enforcement should be deployed when necessary to assist people getting to the help they need. Clearly, structured help at that end of the solution definitely needs an overhauling, one to transition the health care system away from being more of a punitive environment with incarceration than a place one goes to as a foundational step to putting their lives back together. Society clearly suffers from this defect in the system. I doubt there's an addict not dealing with deeper repressed problems.

An overriding issue of concern that I do have is the violence associated with drug trafficking, drug abusive behavior leading to violent criminal behavior to support one's addiction, and, tragically, suicide. Criminalization of addiction does not help. The question regarding impact on tooth health lies in the drug's (or groups of drugs) interaction with the microbiota across the body. If it is adverse, it will invariably show up symptomatically somewhere in the body, and I think one should just count on that happening.

Institutionalization of our food chain

As many societies world wide began to recover in the wake of World War II, international corporations begin to supplant genuine ethnic diets, creating a "civilized" industrialized food chain that has completely remapped what people eat along with their food sources. The result for too many has been dependence on fast food eateries and processed foods, especially younger generations who often lack basic skills to cook fresh wholesome foods. Indeed, as older generations have passed on, so have their recipes, natural food sources, and presence in the kitchen to pass on their culinary skills. Invariably, cook books have become increasingly generic and publishers go with what is considered "politically correct," often playing right into the hands of the industrialists. Grandma's ethnic recipes never stood a chance to get into these books. As an example, my Spanish grandmother's "arroz con pollo" recipe is now lost, and it was incredible — passed down for generations of Castilian Spaniards before her. She died before I could get it from her. I went to my Uncle George (her son) in hopes that he got it, but he didn't, instead sending me a recipe from a cook book that in no way reflected the genuine Spanish tastes and aromas of her dish!

I'm also just old enough to recognize the broader disappearance of regional ethnic dishes. During the late 1950s and early 1960s, on the weekends, my mother and stepfather took me to eat at these very "old fashioned" restaurants in downtown Long Beach, California. They were located inside these ancient (or so they seemed to me), but very stately hotels. Everyone was "dressed to the nines" in their Sunday finest. I still remember the food, and it was good and wholesome. Residents of the hotels, all of them seniors, took their meals here. Music was quiet and pleasant. All the servers were seniors themselves, and "proper" etiquette reigned, both comforting and respectful. And it seemed everyone knew everyone, including my parents. First names were applied only to children, who were scarce or on "good behavior!" Otherwise, it was "Mr. Smith," or "Mrs. Smith," as the case may be, always cordial. Old World gentility, as I think of it. Against this social backdrop came the food on wheeled carts. Fresh steamy dinner rolls, dishes I'd never heard of, everything delicious. All of that is now gone. Buildings demolished, generations passed on. Thinking of that great loss from first hand experience, I'm nearly driven to despair.

In the mid-1970s, I happened upon a festival celebrating Creole, Cajun and other regional ethnic dishes in the bayou country of Alabama. The dishes I sampled were out of this world delicious. But that's all gone now too. The cooks involved were probably in their 80s then, and the venue was on an old slave plantation. You can imagine how that would go over today! But, even then, the spirit of racialist hate handed down from antebellum Deep South was still alive, a reminder that the Civil Rights Movement of the turbulent 1960s was still far from over. Several weeks later, on horseback, I happenstance stumbled upon a huge Ku Klux Klan rally on the rural outskirts of Mobile. Coming from California, I'd never seen anything like it. There were thousands and thousands of hooded Klansmen holding torches on this hill, flickering like a giant swarm of lightening bugs. They would be afraid to show their faces today, as they were then in day light, and with good reason. But they were of an earlier generation too, and, like their Southern culture of ethnic dishes, are also now gone except for small pockets of diehards. How could such good cooking come from such an evil culture of human bondage, arrogance, and racial discrimination and terrorism towards one's fellow human beings?

A word to my vegetarian and vegan friends . . .

I have a number of vegetarian friends and colleagues, most of them right in my own organizations! I have to admit, they are a healthy, fit bunch. But, they also are plagued with tooth and gum problems like everyone else. I can declare 'til the cows come home that we are natural omnivores by virtue of our DNA, but it will only fall on deaf ears. If there is a dietary path around this conundrum, I'm sure they will find it. It's a big enough problem as it is for Paleo researchers to discover exactly what our ancient ancestors ate at the dawn of our species, although they are looking. Even nutritionists today can't agree on what's best. As I wrote in Chapter 2, our teeth, if not our entire bodies, will reveal the truth of the matter. I'll have much more to say on this in the Epilogue, and there's some good news!

§

Chapter Summary

In concluding this chapter, I've tried to point to both concerns and solutions. Our microbiota are at war with us and diet will be an important, if not the most defining, player in bringing peace within our specie's unique biosphere. As you have seen, I don't recommend vitamin supplements or any kind of supplement. This is because I believe we can achieve a balanced diet with healthy teeth through nutritious foods alone. I believe the right choices exist within our current available foods, but will require education and discretion to do what is right with them. Clearly, the path forward on the dietary front will have to be pioneered, as our house is not in good order as it is. Not only is rampant tooth decay evidence of this, so are my fellow human beings whose immune systems and bodies are completely ravaged by cancer, diabetes, Multiple Sclerosis, and other manifestations of WHID. In summary, my dietary recommendations can be condensed to a single concept that transcends the notorious 3 Tooth Truths: Healthy teeth and gums translates to eating a broad range of healthful foods that will deliver the nutrients required to serve our specie's vitality, the same vitality as our ancient ancestors that gave them great teeth. Their DNA lives on in us. There is no escaping our past, and I see no need to either.

Chapter Eight
Exercising for Healthy Teeth!

That's a photo of me trimming a horse in my early 70s. Just look at that healthy, thick flowing white hair, and those muscles – and no steroids either! Well, they didn't arrive by accident. I had to work for them. Male or female, we have to work for nice bodies. You can't have nice, healthy teeth either, living sedentarily on chairs, sofas, and chaise-lounges. We have to work for them. And that means exercising. It's time to get up and suit up. Let's go!

§

I have to admit, including a chapter on "exercising" in a book about dental care, at tooth's surface, seems like a bit of a stretch. But the only stretching that's going to be happening is your body in our Natural Dental Care gym!

The argument for exercising, of course, cuts to the core of common sense in every walk of life. In my profession, natural hoof care (NHC), if you don't exercise and stay in tip top shape, your body will be in big trouble before the sun sets on your first day in the field. In our organization's NHC practitioner training program,[1] most of whom are women, exercising is compulsory and confirmed by our instructors. The exercises we use are largely based on my personal fitness program as an NHC practitioner for over 45 years, which, in turn, was based on my military fitness training in the U.S. Army in the late 1960s and my participation in high school athletics before that.

Meet your gym instructor — Jaime Jackson!

But why exercise specifically to help our teeth? The obvious reason is that as our species transitioned from Paleolithic to Neolithic to Modern lifestyles, the

[1]Institute for the Study of Natural Horse Care Practices (www.ISNHCP.net), a career training program for natural hoof care professionals.

physical rigors of life – requiring strong bodies to survive – declined in our daily lives. The smarter among us realized somewhere along the way that some kind of extracurricular activity – "exercise" – was needed to offset otherwise atrophied muscles from an increasingly sedentary existence. Sedentary lives with such things as obesity, diabetes, cancer, mental illness, and bad teeth. In my day, "gym" classes were compulsory in public schools to get us into some kind of shape, and, hopefully, inculcate in us an awareness of the strategic importance of exercising to optimize good health. Indeed, whereas the concept of extra-curricular "gym behavior" would have made no sense in rigorous Paleolithic lifestyles (although I've little doubt that they had wild tribal dances and recreational activities), we recognize today that our very lives and health are at stake if we don't do something besides sitting, sleeping, walking to the car.

Cooper and other scientists have revealed that our earliest ancestors (*H. sapiens sapiens*) in the Paleolithic Era had somewhat larger mandibles (jaw bones) than we do in the present. No doubt this was an adaptation to masticating raw foods without reliance on the kinds of refined eating utensils we use today. Powerful jaws and strong healthy teeth attached to muscled up bodies simply made sense if one had to chew meat off skeletal parts and then heave or drag the inedible bony parts 50 feet into the local midden site! Today, we reach for the "fork and steak knife" — not just because it is considered "uncivilized" to chew and tear our way through every part of the edible flesh food and raw vegetation — but because too many of us will crack our weakened teeth on the bone. Not to mention all of that hard candy.

Highly refined foods in the wake of the Neolithic revolution probably explains our downsized (called "gracility") and weakened oral cavity, a consequence of the "negative side" of natural selection. Because the shift to a more natural diet will require us to eat more naturally, it makes sense to exercise in such a way as to strengthen the muscles we use to masticate flesh foods that are still attached to bones and as many healthful organic raw vegetables and fruits we can reasonably incorporate.

The good news is that we don't have to target our jaw muscles anymore than any other muscle group. Working all the muscles of our body does this automatically. Our natural dental care exercise program does this very effectively. At the core of the program is the concept that weight-bearing and compression based exercises favor muscular vitality across the entire body, including the muscles of our jaws. This is based on the fact that the gravitational force (g-force) in nature also operates this way on our bodies, and it does it all the time, whether we are consciously aware of it or not. An obvious exception is in outer space, where the stark reality of the "missing" g-force is apparent, but conveys an important message from those among us who have "ventured to where no man or woman has gone before." Astronauts who have lived for weeks or months on end in relative weightlessness in their space ships experience muscular atrophy and fatigue upon reentering the earth's gravitational field. Thus, nature has selected our species to endure this force, and the more active we are in our lives, the stronger we become as a result of opposing it – optimally through weight-bearing and compression type exercises, augmented with a daily walk-jog stroll through our local neighborhood park. The rigorous lifestyles of early *H. sapiens sapiens* would have exaggerated the beneficial effects of the weight bearing force in contrast to the sedentary, sleep-mope-worry wart lifestyles of too many people of all ages today.

The benefits of regular exercising are incalculable. We look better, we feel better, and we are healthier – and our teeth are no exception to the benefits. From optimal vascular circulation to physical strength to resistance to disease, including oral disease, we are doing the right thing to exercise. Indeed, vitality is a natural corollary of exercising and eating a relatively natural diet, regardless of one's age. So, what follows is what I do and recommend. It's probably what most athletes also do regardless of their disciplines, on some level, with specialized adaptations and augmentations based on the specific rigors and requirements of their individual disciplines. Of the latter, I would dismiss those who clandestinely abuse anabolic steroids.

Recommended exercises

2022 marks my 75th year on the planet. Until my 72nd year (switch from practitioner to instructor), I been working under horses since the 1970s and here's some facts you are wise to consider and ponder. First, it was easier for me to trim horses in my 70s than when I was in my twenties! Unless the weather is very hot, I rarely sweat; sweating is almost against my religion, meaning if I sweat I'm not being very efficient. Most of the time, when I've finished trimming a horse, I feel invigorated but I also feel like I've not done anything, when actually I have done a lot. Watching me work, it appears also that I'm not straining myself breathlessly.

But if I were to put you in the same situation, you would be pouring sweat and crying in the night for opioids! So, it's not that I'm not working hard, it's that I'm *conditioned* to the work. But none of this would be possible if I didn't exercise the way I do. It's a fact.

Physical conditioning exercises are the "Golden Rule" to follow if you expect to do my work, or any extraordinary physical labor, well into your 70s and beyond without pain and regret. I warn our students, if you think you're going to muscle your way through this without conditioning exercises, you will regret it one day with back pain and probably back surgery (that will probably make matters worse). Or you'll just give up from muscle fatigue and shortness of breath half way through the first hoof, unwittingly saving your back from certain catastrophe. We get those too, and our training program weeds them out for their own good.

Meet your NDC exercise instructor – the author!

In fact, I consider myself an expert at exercising. My expertise began when I was a teenager with Osgood-Schlatter disease — a painful inflammation in the knee area that occurs during young adolescence, particularly in males. The doctor took me out of regular "PE" at school, and I was put in what was derisively called the "ortho squad" by "normal" students who didn't hesitate to joke about us when gym teachers weren't nearby to deflect their derogatory comments and defend us. There were about 12 of us — all suffering from either debilitating congenital diseases or, as was my case, temporary chronic inflammations, that prevented us from walking or functioning like everyone else. Several suffered from things like polio and heart ailments, and life for them would always be a struggle and under the constant threat of an early death. I felt compassion for them, and even inspiration as they did whatever our gym teacher, Mr. Stratton, could facilitate. In no time, we became a "team," helping each other and taking pride in our "squad's" accomplishments. I always look back on those days as among my very best.

In addition to my knee problems, I wasn't doing too well on other fronts either. A really pretty girl I knew, whose attentions were much sought after by other boys, and who, for reasons I could never understand, always wanted to walk to and from school with me, even though she ridiculed me one day in front of others, "Jaime, you have no brains nor brawn." But the fact is, she was right. I was skinny and, admittedly, a vocal trouble maker in class - in fact, I was deservedly flunking (and did flunk) the 8th grade. Plus, at the beginning, being condemned to the ortho squad didn't help any for my ego. I think Stratton, a crew cut WWII veteran as I recall, saw my despair and, on the first day in the ortho's "segregated" workout room, he said, "Jaime, you are going to work with the 'weights' and exercises that you can do." I had no choice but to submit.

What he put me through became the foundation I use pretty much to this day, with some adaptations I came up with for my work with horses. By the end of the school year, everyone I knew began to notice that I was looking different. Then one day, in the boy's locker room, where there was this giant white billboard with neatly printed black lettering that listed all the school's athletic records, my name was added for setting the school records for push-ups and chin-ups with Mr. Stratton doing the official counts in the ortho gym. This brought me to the attention of the school's gymnastics' team coach, Mr. Romo, and also the nearby high school's coach, Mr. Bellmar. I never returned to regular PE, but was shunted into competitive gymnastics where I excelled in the 20 ft. rope climb event, and won medals as a ring man.

I am forever indebted to this string of teachers who taught me how to develop my body and mind, and later, fellow gymnasts who with similar training, inspired me to learn what they had achieved. It also helped clear my head, and at summer school that year, the discipline of athleticism naturally transferred to my academic side with more than passing scores. The experience laid down the foundation for me to become a questioning critical thinker.

Iron Cross, Jackson.

So, what follows is an exercise program that works, costs nothing more than a few dollars, and 20 to 30 minutes of your time a day. What you can get out of it is a strong, healthy body, and — coupled to our natural dental care program — healthy strong teeth. Chances are you are already athletic and in top shape. That's great, but read the following anyway, *just in case.*

20 minute morning exercise routine

I'm going to tell you what I do now to keep in excellent shape, six days a week (I take Sunday off). It's how I start every day, first thing every morning at 5 am for no more than 20 minutes, followed afterwards by a walk-jog on the side of a hill. For the most part, I've been doing these exercises (slightly modified) since the days of the ortho squad when I was 13 going on 14. You'll need some dumbbells, hand grips, and a space a little bigger than the length of your body by the width of your arms stretched out to the side. I use a bedroom rug. You don't need to go to a gym for any of this. Your body and mind becomes your gym. From there, it all starts with the concept of *progressive development*. Here's a related quote from our ISNHCP exercise training program explaining what that is:

> The following are daily exercises; typically, they are grouped in "sets". Each set includes "repetitions" (reps) of the particular exercise. It is also important to take days off to rest your body, during which no exercises are recommended. Here's why: multiple sets gradually develop the muscles of your body for hard work (like trimming horses). At first, you will do fewer reps; but as the weeks and months go by, you will automatically be able to increase the number you can do. This ia *progressive development*. It applies equally to physical tasks unrelated to your exercise routine. For example, "You will only be able to do a few hooves at a time, maybe only one hoof. Then, through progressive development, you will be able to do two hooves without stopping, then three, and, one day, the entire horse, and then many horses!" At first, your mind will be saying, "I can't do this". But, once more, through progressive development, your unconscious mind will begin to "take over" and say, "Now I can do this." And you will!

Muscle tone, balance and relaxation

Notice in the photo on the previous page that I'm actually smiling during the "Iron Cross" on the still rings during a competition. It may come as a surprise, but I'm very balanced and *relaxed* — and I was required by the competition rules to hold this position for 4 to 5 seconds before transitioning out. During some routines, I would transform the cross into its variations and sustain them collectively for 8 to 12 seconds. The physical and mental state of being relaxed in the middle of a power move is called "collection," and represents the convergence of well-developed *muscle tone*, physical *balance* relative to the g-force (very important!), and a mental state of calm *relaxation*. Of course, I couldn't do any of that at the beginning, but through strategically planned progressive development I could. It even surprised me at the beginning! During one of the high school team's many practice workouts, I carried on a brief dialogue with one of the school's football players, a friend of mine who enjoyed watching the ring work, in particular the "power moves." Half again my size and powerfully built, he always wanted to give it a try himself. We laughed together, as it was all he could do but just hang like a sack of potatoes from the rings, unable to do anything else but let go from aching hands! We each followed completely different training paths to athletic excellence, but both were based entirely on progressive development (and hard work!). Collection is one of the many benefits of following my time-tested routine that follows, serving optimal health including our teeth. I sip a small bottle of chilled clean water throughout the exercise routine — just like I did as a gymnast during workouts and competition events.

Stretching

This is such an individualized thing to do, you'll have to figure this one out for yourself. I personally do mine standing up, bending this way and that depending on how I feel. I don't do warm-up exercises because they don't make any sense to me. The late Jack LaLanne (discussed a bit later in this chapter), a fitness expert who also influenced me, was dismissive altogether of "warm up," calling it "shtick."

Push-ups

This is truly the great muscle builder that covers the entire body (*next page*). Some very impressive athletes do nothing but push-ups for their exercising. World records are astonishing, in the tens of thousands! Record holders are unbelievable: over 3,000 in one hour, 46,000 in 24 hours. Some do them on the backs of their hands! Others using just one arm. Some just one finger! The most knuckle push-ups in one hour by a woman is 1,206 by Australian Eva Clarke (*top left, next page*) in 2014! Whoa! We won't go to such extremes, and I do hope these competitors aren't using steroids! I think a good number to shoot for is 100 push-ups in four sets of 25, spread out between the other exercises (e.g., sit-ups). You'll find that the sets get easier as time goes by as muscle mass develops and conditions. It's amazing what the human body can do, when we try.

You will probably feel "muscle burn" as you fatigue. This is due to glucose

There are many variations of the push-up, one of the great "compression" based exercises for muscle development. Find your way to one or more of them. At age 70, I do 100 as part of my 20 minutes exercise regimen, but am conditioned to do 1,000 in about 6 hours. Here I am at 17 doing a push-up into a "planche" with my legs off the ground on my way to a handstand — which I also did on the rings! If you're new to the push-up, they're tough at first, but incredibly easy later if you follow the principle of *progressive development* — you can do it!

breakdown by the body resulting in lactic acidosis.[1] If this occurs, simply stop, walk around or stretch a moment, rest, and then resume. It's perfectly natural and nature is just warning you to rest a few moments before continuing; when the burn subsides, continue. It happened to me all the time as a gymnast and happens to me everyday that I exercise as I reach my self-imposed limits. No big deal! Also, as with any compression exercise, breathe in as you let yourself down, and blow out as you push upwards; whatever, just get enough oxygen to continue. Shoot for one push-up per second.

If you can't do a push-up with your body levered straight from head to toe, do them on your knees. If you can only do one or two, that's where you start. Over time, possibly many months, you will eventually graduate to knees off the ground. Obviously, you can't do a hundred at the onset (i.e., 5 sets of 20), so your goal will be whatever you are capable of doing over your 20 minute regime. I do 20 in about 20 seconds (50/minute) — the world record is 134/minute done on the back of the hands!

So, do your push-ups in sets of whatever you are able to achieve at first, on your knees if needed. If you simply can't do push-ups but on your knees, that's okay, do them like that because it's still very effective. According to a study published in the *Journal of Strength and Conditioning Research*, the test subjects supported with their hands, on average, 69.16% of their body mass in the up position, and 75.04% in the down position during the traditional push-ups. In modified push-ups, where knees are used as the pivot point, subjects supported 53.56% and 61.80% of their body mass in up and down positions, respectively.[2] Nature will let you know eventually if you can do them levered (hands only). If 100 proves to be beyond your reach (five sets of 20 = 100/20 minutes), then pick a lower number and set that as a goal (e.g., five sets of 8 = 40/20 min).

I remember when I did 1,000 in early 2017 over one full morning (about four hours, after talking to Jack LaLanne for advice.[3] But in 2018 through 2021, I

---

[1] There is some debate concerning the actual effects of muscle "burn" and lactic acid. In the complex biochemistry of muscle fatigue, lactic acid is produced, but it's presence is ephemeral as part of a larger chain reaction. But whatever acid source, burn translates to "pain," and we get the message! An interesting read that discusses the issue: Lindinger, M. I. (2004). "Applying physicochemical principles to skeletal muscle acid-base status". *American Journal of Physiology*. 289 (3): R890–94.

[2] Suprak, David N; Dawes, Jay; Stephenson, Mark D (February 2011). "The Effect of Position on the Percentage of Body Mass Supported During Traditional and Modified Push-up Variants". Journal of Strength and Conditioning Research. 25 (2): 497–503.

[3] He told me, "Do as many as you can until exhausted, rest, and then continue immediately until exhausted again, repeating, etc."

upped that to 500 in just 25 minutes on odd days and 150 in less than 10 minutes on even days, approximately 10,000/month. When I turned 75, I lowered it back down to 200 in 4 sets of 50. (I plan to hold that number into my 90s.) In addition to push ups, I also do a whole lot of dumbbell exercises, sit ups, and other things I'll discuss in a minute. Once again, the principle of progressive development is the key to reaching you push up goal. But then, you're not going to trim horses, or try to set a world record for push-ups are you? Or are you? Do what seems right for you.

Squats

Another great exercise, I do only one set of ten and I stop bending before my thighs are parallel with my knees, called high squats. Some experts claim "low squats" – below the parallel – put the lumber spine and knees at risk of injury over the long term. Well, at 55 plus years of doing them this way, I concur with other trainers who believe that high squats are one of the best exercises for safely developing muscles and strength. Likely, the problems with low squats occurred because of adding heavy (barbell) weights to the exercise and overloading the body. But since we don't do that, and there's no need to, our high squats are perfectly safe. My "rule" is to always have my feet in view during the squat and keep my thighs higher than my knees. Also, keep your heels always at shoulder width, which facilitates an optimal "platform" for total body balance.

High squat

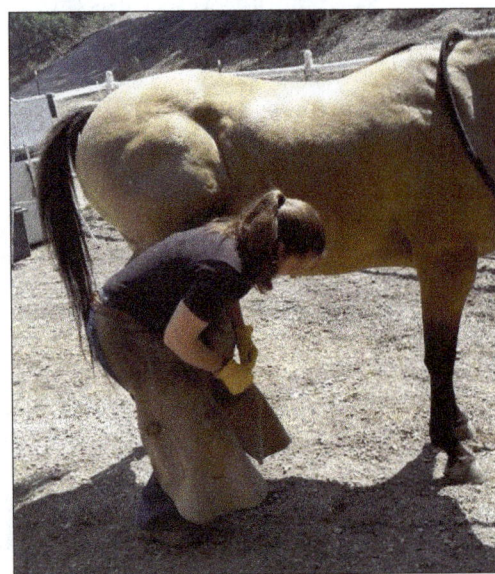

ISNHCP clinician takes the right hind hoof in the "deep n low" squat position. Not only a great exercise for developing strong legs and buttocks, as you can see, it has the potential for many applications in life that relieves the lower back of unnecessary stress and strain. Key to my ability to do this work in my 70s!

I conclude my squat set by holding the final one for 60 seconds (or until lactosis forces me up!). I do this by bracing both of my forearms near the elbows on my thighs just above the knees; I usually alternate by substituting one or both hands. I call this the "deep and low" position for trim work quite literally under the horse, but it's particularly useful when lifting any heavier objects. Squats bring to bear our powerful hip and leg muscles, rather than bending over at the waist to lift which is a sure way to injure your lower back (sciatica).

Jack LaLanne — one of my exercise heroes!

So, they are a very important part of our training because of their diverse applications (I'm sure not for trimming horses in your case!). Surprisingly, stability and proficiency in deep and low enables one to actually relax and do chores (whatever they might be) at ground level. We will draw upon deep and low again and again during the dumbbell exercises coming up.

Sit-ups

Like push-ups and squats, there's many ways to do these. My preference is to lay on my back with knees drawn up and hands clasped behind my neck, then simply tighten my stomach muscles, but always keeping flat on my back (*next page*). At one point I was doing 400, but that turned out to be an unnecessary number of reps. I do 100 now and that does it for my purposes. Another effective way to develop the stomach muscles is taking the "dying cockroach" position we we're forced to take (as punishment) in Army basic training (*next page*). There are actually a number of variations of this exercise, and I incorporate half a dozen into my routine. Find your way to as many as you can think up and that work for you. Be creative with all your exercises!

I complete the sit-up exercises by balancing myself on my hips like the late famous exercise aficionado Jack LaLanne and his wife Elaine are demonstrating above. I've included a second stretching exercise featuring LaLanne which is also part of my "stretching" exercise routine. LaLanne and his wife lived just north of Lompoc in Moro Bay, before he passed away at age 96 in 2011. He exercised daily

"Dying is easy. Living is a pain in the butt. It's like an athletic event. You've got to train for it. You've got to eat right, you've got to exercise, your health account, your bank account, they're the same thing. The more you put in, the more you can take out. Exercise is king and nutrition is queen: together, you have a kingdom."
Jack LaLanne

Recommended sit-up position.
To avoid injury and sciatica, my recommendation is to keep the back flat on the supporting surface. From this position there are a surprising number of variations to go to. I've included a few here. I do all of these and more in my morning exercise. The bottom image at left is basically the "dying cockroach" position the Army drill instructors put us in daily for both "rewards and punishment." Work your way to this one, although few of us will be able to be as exemplary as the subject here. Just keep your legs up as high as you can. Hold it for a few seconds, then down with bent legs to rest, then up gain, etc. Progressive development will get you there. Patience!

for several hours until his very last day. My Uncle George also exercised regularly and ate well, but other than short walks, did not exercise in the gym or in his home in his later years. George died four years after LaLanne at age 101. In contrast to my uncle, LaLanne did amazing feats of strength, including 1,033 push-ups in 23 minutes at age 42 on TV in 1956! In 1984, at age 70 — handcuffed, shackled, and fighting strong winds and currents — he towed 70 rowboats, one with several persons aboard, one mile across the Long Beach Harbor in Southern California! Holy smokes! Not surprising, LaLanne was an inspiration to a young Arnold Schwarzenegger just arrived in the U.S. from Austria.

A fleeting thought comes to mind about these compressional exercises. At their center is the understanding that we are engaging in contraposing forces that bring muscle groups against muscles groups (e.g., lower body muscles opposing upper body muscles as in sit ups). This tension, I mentioned earlier, endows our bodies with "collection." But, there is more to it. By strengthening our muscle groups, the major muscle players, in turn, act upon tendons, which then act upon skeletal structures — including their joints. Thus, exercise, in building strong muscles, also protects our bones and joints. Add to this optimal circulation and we can say that muscles also contribute to the "nutrifying" of these codependent structures. In this interpretation, it is the muscle groups, stimulated by the nervous system, that keeps us erect and functional. If there was ever a message to stay healthy and strong, it would be this "interconnectedness" driven by muscle building and toning through progressive development of our muscles. Our Paleolithic ancestors must have exemplified this, to say the least!

Hand grippers

I use these hand instruments primarily to simulate nipper work, which requires hand strength. Some have adjustable spring tensioners — I use the $6 Wal-Mart gripper which works fine (*left*). Make sure the tension spring is something you can readily squeeze or you'll be complaining of carpal tunnel syndrome before long. I do fifty reps alternating squeezes between my left and right hands, then both hands simultaneously — all the while doing squats (my way of increasing the number of squat reps but spreading them out over the entire workout). When I'm done with these 50 reps, I finish up in the deep and low position, where I count to 60 before vacating the position.

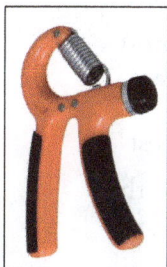

Dumbbells

This is the last group of exercises I do. Their purpose is to strengthen and condition the arm, shoulder, and chest muscles. They really give you that "cut" look, too! The first question is, what weight dumbbell does one use? The rule I go by is select the maximum weight you can comfortably and safely control through an entire set of reps. Follow the images (#1 thru #5) as I explain what to do.

#1 demonstrates the alternating bicep curl; #2, the two arm curl. Notice that both subjects have their wrist/palms facing towards their bodies in the pre-lift, lower position. This is the most common position for most people based on human anatomy. Confirm that is applies to you as follows: without holding the weights, let both of your arms and hands hang loosely down at rest at your sides. Typically, the palms are facing towards your body and slightly rearward. If this is the case, then this is their natural starting position for holding the weights.

Next, raise the dumbbells so that so your palms are facing towards your body as the two subjects are doing. Depending on your upper arm conformation and musculature development, the dumbbells will either terminate at a slightly oblique (#1) or level (#2) angle. Lower them back down. In either case, you'll notice that as you raise and lower your arms, they rotate away from and towards their resting positions, respectively.

You may be asking why I'm not recommending barbells or weight machines. Barbells take up more space and are more limited than dumbbells in how weight can be manipulated with our arms and hands. The same goes with weight machines. For our purposes, neither are needed anyway. Arnold Schwarzenegger wrote in one of his body building books that he didn't like using

weight machines because they cramped the natural movements of his body, including his arms, wrists, and hands. I agree. I also find barbells very restrictive for doing curls, and wouldn't recommend them either for exercises #1 and #2.

Okay, as an additional exercise, go back to your basic push-up position and notice how you naturally position your hands on the floor. You'll probably find them rotated slightly inward, depending on your unique conformation. If so, that's your starting and ending position. It should feel natural and comfortable. If you're feeling any undue torsion (twist) on your wrists, you've not found your natural position. Now, go back and look at my wrist position in the floor exercise photo (page 105) — try to figure that out! As I complete my press to the handstand, my wrists began to rotate inward until I'm back at the basic push-up hand position (ironically, the "rest position"). All this detail I had to scrutinize in order to do these power moves. And in each instance, I was seeking out my natural wrist, hand, arm and body positions and their corresponding trajectories. One doesn't just do these things, one figures them out. And, I would add, your overall sense of balance is also germane — meaning finding your center of gravity.

Moving to #3 (*above*), the subjects demonstrate two dumbbell presses. In addition the

same presses can be done alternating the left and right dumb-bells. I do both. As is always the case when beginning any of these exercises, examine any tendency for your wrist to rotate and follow that movement — do this first without the dumb-bell to confirm, then with them. I do 20 alternate reps first. Before I do the two arm presses, I switch to #4 (*right, below*) doing 20 reps, which adds further dimension to upper body strength and muscle definition, but without fatiguing the muscle groups used in #3. Back to #3, I'll do two sets of 10 two arm presses, then finish up with #4 (*right, above*) doing 10 reps.

This brings us to #5 (*above*), which I used to develop my chest muscles for the still rings, facilitating the Iron Cross and other power moves. But be careful with this exercise! I do mine on the floor rather than on a workout bench (as is commonly done and depicted here), because there is less chance of the dumbbells pulling my arms down out of con-trol. And because I also take the opportunity to do other dumbbell presses while on my back with intermittent sits ups. You can do this exercise with bent or straight arms. I do mine straight armed from habit as a gymnast. To avoid injury to your shoulder and arms, start with a weight that you can easily control, and do fewer reps to begin, and build from there. I do two sets of 10 lifts, with a short break in between sets. This is a great power builder, but stay within your

#4 — There are different variations that are possible here. These are the ones I do, (*top*) pulling upwards from the front and (*below*) from the side. In either case the movement is supported by bending forward slightly, which activates the abdomi-nal muscles, which, in turn brings the full power of the lower and up-per body muscle groups. I will do one set alternating the dumbbells, the other at the same time.

body's limits.

I finish by doing 40 push-ups, which, you may find as a surprise, are easier to do than the first set of half that number. I'm conditioned to do 100 consecutive push-ups, but my routine is designed to develop the muscles for shorter work spans, which is more in sync working under the horse. I saw a woman years ago who could bend a thick iron horseshoe with her "gloved" hands — her chest muscles being the "key" to doing this. This can also be achieved by about 30% of our fellow (gifted) human beings through what I call "deep rapport" concentration. The horseshoe literally bends as easily as flexing a soft rod of rubber.

Ladder climb

Here's another good one, if you don't have ready access to stairs or a hillside. If you have space, get an 8 ft. ladder and climb up (for stability, just part way up) and down, repeating to create suitable reps based on progressive development. If you over do it, you'll have aching arches — they'll recover, but in the meantime not fun! This exercise will strengthen the arches of your feet, the calves, thighs, and hips.

Start or end of the day walk/jog/bike

The workout routine above keeps you in that 8 x 8 ft. space. To balance that, I walk/jog a mile once a day, and may ride my bike the distance too. If you're going to walk, walk like a soldier on a forced march, meaning dig your heels in and really move along. Not a time for sauntering and lollygagging! If there's a nearby hill or other natural incline, preferably steep, incorporate that too — and you won't need the ladder. But whether walking, jogging, biking, or climbing a ladder, this is no time for procrastination, get off the sofa and get going!

§

Summary

In all of these exercises, I don't stop to take lengthy breaks, just long enough to sip water, a few seconds, which I do throughout the routine — and as I do with trimming horses. So, it's pretty much non-stop. Jack LaLanne advocated for the continuous (non-stop), fluid integration of exercises, which I obviously fully embrace.

It may be that in the course of doing your exercises, you experienced pain, or perhaps you arrived at this natural dental care program with existing pain issues due to trauma injuries such as repetitive-use breakdown (e.g., carpal tunnel syndrome), lower back problems (e.g., sciatica), and upper back problems (e.g., rotator cuff tear). My cardinal rule in life is to "treat" pain through exercise, based on the premise that the g-force is a healing force. As in dentistry, I am opposed to invasive procedures when nature provides a remedy that channels effectively through our body's immune system, diet and native biomechanics. Accordingly, using strategic periods of rest (and sleep), and do those segments of the exercise regimen that do not cause pain, thereby bypassing trouble spots. Trouble spots, in turn, are approached by testing with minor weight bearing or compressional forces. Invariably, trouble spots are mitigated over time when supported by those exercise segments that are actionable sans pain. Technically, we do this by ceasing a particular move; moving to a variation of the exercise (I do this often at the first signal of hypersensitivity as a preventive measure); bypassing the problematic exercise altogether; giving the body time to make its adjustments ("internal balance") to restore itself; adjusting hand, feet, joint and upper-lower positions. In short, take control of each movement, your sense of physical and mental balance, and your physical limitations through progressive development. It's your ship — command it. Be the captain of your body, including your teeth!

I listen to New Age music to go with my work flow, as it helps mitigate any mental or environ-

mental noise that might disturb the routine and my deeper mental purpose to heal horses and heal my teeth. Exercising for me is a Zen — meditative — experience, as it should be, because vitality invites our deepest thoughts and understanding that "all things are connected" and that the universe is a healing force. We are healing and guarding our teeth from the brutality of a dental industry that only wants to remove them from us. Exercising is an important way to fight back.

Your first assignment in this chapter is to move quickly to create your 20 minute exercise regimen and then initiate your one mile walk/jog/bike/ladder routines.

To our vitality!

Chapter Nine

Our Natural Dental Care Resources

To go forward with this natural dental care program, you will need some core products to make the most out of it. It's so important that you not use the healing agents with ADA recommended toothpastes and rinses, as I explained in Chapter 6. If you use those at all, you will neutralize the effects of the healing agents on your oral microbiota because those products are intended to kill them off, not bring them into balance. So, if you do that, you're basically back in the chair, and the program won't help you. If you decide to retain your relationship with your dentist, and combine plaque removal treatments with our program, your ship is really sunk. I'm very dubious that they will support you in this new way, for the simple reason that "natural healing" isn't something that they're trained to know anything about or to do. Be forewarned: conventional dental care is all about "search and destroy" — killing oral microbiota (biofilm), removing tartar, and grinding away tooth structure. Natural dental care is exactly the opposite: balancing the microbiota colonies (biofilm), building strong tartar, and keeping all of our teeth intact. Finally, to keep WHID in check, you must take a critical eye to a reasonably natural diet discussed in Chapter 7, and, of course, the exercise program in Chapter 8. Just about everything below can be ordered at WalMart!

Toothbrush

I recommend using a brush with the least aggressive bristle you can find. There are many synthetic and natural fiber brushes (animal hair and plant based) to choose from. Go for the "soft bristle" versions. Most manufacturers, conventionally or naturally oriented, are still working from the premise of getting rid of plaque as part of oral care, rather than using the toothbrush as explained in Chapter 6 to lightly clean the teeth, gums, and tongue. A trip to a health food store to see what they have is recommended, but you may find a conventional soft bristle brush works as well or better for you. Try both types and see what suit you best.

Water flossing

To augment brushing, I highly recommend any of the "water flossing" machines such as the Waterpik Ultra Countertop Water Flosser WP Series (100, 112, etc.) if you have sensitive gums. The water pressure streaming through the pik can be adjusted to your sensitivity. If sensitiviy isn't a problem, then opt for conventional dental floss recommended below.[1]

String (filament) flossing

As I wrote in Chapter 6 ("Flossing"), I see flossing as a way to remove aggravating food stuck between the teeth that may be aggravating the gums. There are many brands of filament flosses to choose from. I recommend the "Oral-B Glide Pro Health Deep Clean Floss (Cool Mint)" because its narrow string gauge enables one to work past fillings and crowns without dislodging them, and because it is ineffective in damaging formative plaque.[2]

Bottle warmer

Using the Waterpik with cold water can be brutal with those having sensitive teeth. To solve this problem, try a baby bottle warmer. Quick and simple to use for warming your filtered clean water.

Tooth powders, pastes and rinses

It is regrettable that both conventional and "natural alternative" powders, pastes, and rinses/mouthwashes continue to signal anti-plaque objectives with their products. I advise going the natural alternative route if they do not include Fluorides, but with this caveat: Because the powders and pastes contain corrosive abrasives as identified in Chapter 6 (please review that information), they can't be used during brushing or they promise to weaken or destroy the biofilm–tartar platform, thus opening the enamel to invasion by *S. mutans* and other aciduric bacteria. But if their tooth powder or paste is soluble in warm water, they could be used as a rinse that would be harmless to the biofilm

Waterpik Ultra Countertop Water Flosser WP-112

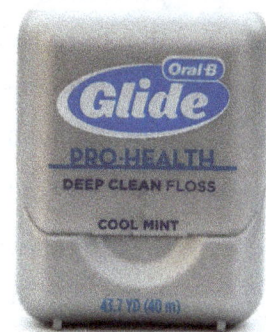

A great product, but careful not to misuse it!

"First Years Quick Serve" bottle warmer

[1] Evidence is strong for "water jet" benefits, even though researchers are "stuck" on plaque removal as one of the effects, which I ignore because it is clear that targeted biofilm fails to obstruct tartar formation:

[2] Berchier CE, Slot DE, Haps S, van der Weijden GA (2008). "The efficacy of dental floss in addition to a toothbrush on plaque and parameters of gingival inflammation: a systematic review." International Journal of Dental Hygiene. 6: 265–279. Citing the authors: "The dental professional should determine, on an individual patient basis, whether high-quality flossing is an achievable goal. In light of the results of this comprehensive literature search and critical analysis, it is concluded that a routine instruction to use floss is not supported by scientific evidence."

because they are not being used abrasively with the tooth brush. Whatever value the non-abrasive ingredients have to offer would be delivered to the oral cavity. Brushing is then done with warm water only and flossing.

Natural Dental Care (NDC)

The purpose of brushing and flossing, according to conventional dental industry mantra, is to clean one's teeth and gums of food debris. aciduric bacteria, bacterial biofilm, and tartar (calculus). Bear in mind, however, that Paleolithic evidence based on tartar composition suggests strongly that dental industry regimes we know today didn't exist then. Well-nourished biofilm and calcified tartar was the order of the day. Limited, but arguably significant contemporary anecdotal evidence by people who do not brush their teeth at all and have no dental caries and gum diseases, point directly at the Paleolithic model. In both cases, we are looking at WHID free diets as the only plausible explanation. It appears, and it is no surprise, there is no genuine interest at this point in time in pursuing this explanation by today's dental science sector. As I've shown, with Cooper's exception, research is constrained by flawed, or if not that, biased Neolithic models that serve research grant funding and market profiteering that is deeply and successfully rooted in WHID.

In following the logic of the Paleolithic model, NDC necessarily "ignores all pathology" in the quest to treat WHID — of which oral disease is a symptom, not the cause — probiotically through nutrition. There is an inherent paradox in this strategy: How can one ignore dental pathology when it is so manifestly ubiquitous?

When I say, "ignore all pathology," what I mean is do for our teeth what is natural for our species. Because when we do — as I have learned with treating horses in concert with their specie's ancient adaptation — pathology invariably wanes in the wake of natural vitality. In fact, this is how our bodies heal themselves, and in this our native microbial populations are key players in our healing immune systems. How can anyone deny that diet and exercise are also not key players? Going one step further, this is not to declare that conventional dental care and NDC should have no role in this healing together, because they should. But both should unite to serve nature's healing powers, not obstruct them, building a bridge between nature and science to the benefit of our bodies, including our teeth.

What is above all else in clarity is that toxicity in the oral cavity has given us tooth decay and gum disease. It has become so bad, that our immune systems cave before the onslaught of a disgruntled and now dangerous oral microbiota. On the one side, conventional dentistry deals with it through "search and destroy" weaponry and tactics to wipe out the invasive bacteria. But this has failed, creating a pathway from cavity fillings, to crowns, to implants, to no teeth at all and dentures. NDC, seeing this failure, points to the Paleolithic model and recognizes that the central problem is not the symptom of dental disease, but WHID. Ruling out genetics, ruling in exercise, we are left with the missing piece of the puzzle: Diet. We must, NDC declares, bring a reasonably natural diet to our teeth. Because it is in the oral cavity that digestion begins. Logically, if we assert dietary changes there, they will soon follow in the lower intestine, and back from there to every part of our bodies, including the roots of our teeth.

I have scanned the marketplace for "natural" non-abrasive toothpastes and powders and have not found a single product. I have discovered a promising paste, however, called **Oralive,**[1] a product of a company called Ascended Health. They appear to be firm believers and practitioners of New Age Earth-Science-Spirit medicine. Although they espouse the typical "anti-microbial" and "anti-plaque" concerns, they do so within the context of "balanced microbial communities" as I do.

Oralive it is made by hand in small batches and comes with the recommendation to swallow it for its nutritional value (absorbed through the oral mucosa), calling it a "super food." Attractive also is that Oralive is available *unsweetened* along with a sweetened version using xylitol, common in many natural pastes used by competitors in the health food industry. I have used it without issue, and reviews for it are similar — but scarce. Because it is edible and contains far more naturally occurring botanical and mineral ingredients than any other product I've found, I will use it as an example on how to use such a product with minimal to no adverse effect on the biofilm and tartar scaffold — again, providing the toothbrush is not deployed when using it. I do look forward to the development and availability of edible toothpastes that support a healthy biofilm colonization, are non-abrasive, and are naturally inhospitable to WHID.

"We believe in effective natural methods to treat chronic illness, not surgery. Our products are designed to treat the underlying infection or abscess and keep it infection-free, naturally. No added Flouride, SDS (soap that makes toothpastes foam) or chemical preservatives."
— Ascended Health

[1]https://www.ascendedhealth.com/gum-disease/unsweetened_oralive.htm#!

Worth noting from Ascended Health's website:

> Founded by a microbiologist with over 30 years experience researching
> how indigenous foods and their ecology of microflora help stave off disease,
> detoxify the body, and increase life span, Ascended Health is a trans-
> denominational, health education and faith-based organization dedicated to
> creating a unique range of effective, all-natural organic products specifically
> designed to help our bodies heal themselves.
>
> Predicated on the belief that Nature/Universal Source has provided us
> with everything we need for optimal health, vitality and well-being, we spe-
> cialize in developing effective plant-based extracts into a bio-intelligent for-
> mat that is eco-harmonic to both the body and spirit through the processing
> of microbes from longevity zones around the world.

Not surprising, this type of New Age language invites an indictment of "pseudoscience"
from legions of the disgruntled naysayers. On the other hand, the science of natural horse
care that I practice based on the adaptation of *Equus ferus ferus*, the ancient antecedent of
the modern horses, has also been targeted as pseudoscience. Fortunately, we still live in a
world where people can — as our own ancient Paleolithic ancestors undoubtedly once
did — think and decide for themselves what is and what isn't "the real thing."

Directions for using Oralive as part of my NDC regime:

A. **Brushing.** Using only warm, clean water, brush your teeth and gums. Floss if
necessary to dislodge aggravating food particles stuck between the teeth. Brush
lightly with a soft bristle brush. Bristles can be softened further by immersing
them in warm water. 30 to 45 seconds should be more than enough brushing
time, just enough to remove any obvious surface food debris. Finish by rinsing
and expectorating with warm, clean water.

  ◊ The warm water rinse is not a bad idea following any "unauthorized" des-
  serts or sweet fixes. Sugar is soluble in warm water so "post-sweet treat"
  rinsing seems logical. But since you are also feeding potential bacterial
  colonies conducive to WHID in the lower intestines, you are duly forewarned of the consequences of conducting such clandestine sweet tooth
  forays a bit too often!

B. **Oral rinse ("pulling") using "Unsweetened Regular Strength Oralive."**

  1. Place a small dab ("the size of a pea") on your tongue. This will naturally
  stimulate your saliva to activate healthful digestive enzymes.

  2. Swish the salivated product between your teeth and gums for 30 seconds
  up to 5 minutes.

  3. After swishing, expectorate — or swallow it for "maximum nutritional

benefit," according to Ascended Health. Do not rinse your mouth with water, as the residual formula and salivation will continue to mediate your oral bacteria.

4. Repeat up to 3 times per day.

Oralive ingredients: Volcanic earth minerals, Vitamin C, Camu Camu, Amla, Calcium, Nettle, Folic Acid, CoQ 10, Marine Phytoplankton, Astaxanthin, Blue Green Algae, Pau D'Arco, Aloe Vera, Noni, Peppermint, Melissa, Arnica, Licorice, Mint, Tumeric, Holy Basil, Ginger, Essential Oils of Neroli, Jasmine, Spikenard, Rose.

Cosmetic concerns

I caution against using artificial tooth whitening products for cosmetic purposes. Tooth enamel is somewhat translucent and naturally varies in color from light yellow to grayish white. Enamel's whiteness comes from hardened calcium phosphate, but that is tempered by the understory contents of the supporting dentin. But how white your teeth are going to be will also depend on the tartar composition (hardened dental plaque) coating your teeth. Tartar is composed of both inorganic (minerals) and — we recall from Cooper — organic (cellular) substances. Although up to 60% of the tartar is made up of minerals, some of which impart whiteness, bacterial cells and other organic materials derived from foods we eat, environmental debris, and many other things, are also embedded in healthy calculus. Opting for tooth whiteness in lieu of nature's protective tartar barrier is contraindicated by NDC.

§

Chapter Summary

In using these dental tools and products, I think it's important to remember that we're not declaring war — specifically, a war of eradication — against mutans streptococci and other bacteria. They're all meant to be there. Our goal is to reinforce the normal mechanisms nature embedded in our body's digestive system to moderate these oral bacteria and bring them into harmony with microbiota elsewhere in our body. Without them, we cannot build strong teeth. Abuse them, and they will inevitably destroy our teeth as sure as the setting sun will come tonight and every night. We've learned that using powerful chemicals in conventional toothpastes and rinses temporarily wipe out these and other bacteria. But that the aciduric *S. mutans will* recover first as nature intends and has built firmly into our

DNA. Other mediating bacteria recover too late — not even copious salivation can save the day! All microbiotic parties need to be front and present!

Once a healthy biofilm forms and begins its formation into tartar, our job is to use the recommended healing paste and gentle brushing with soft bristles to lightly exfoliate any loose tartar debris, followed by the nutritional paste, and nothing more. Brushing, therefore, is minimal as explained in the instructions. But why brush at all, if our Paleolithic ancestors didn't? Basically, to finish the job that our gracility and diets lacking roughage from not eating enough raw foods have converged to fail us.

From the archeological record it is clear that, in our beginnings as a species, nature never really intended us to brush our teeth with tools and machines, eat processed foods, or enter the dentist's chair to remove plaque. Indeed, researchers have shown us that dental calculus was replete with food debris in our Paleolithic ancestors, as were the teeth of indigenous peoples living so remotely in the "hinterlands" of modern life as to be left virtually unscathed by civilization. But, there's no doubt either that the natural diets of "primitive" peoples required strong teeth to eat more raw foods and fewer processed sweets than people do today with our industrialized food chain. The absence of free sugars from their diets also meant that they would have instinctively sought out and eaten wild fruits and other fructan rich plant life to satisfy their crave for energy to burn in what can only have been a very rigorous lifestyle. The amount of naturally "reduced" sucrose from these Paleolithic sources not only met their energy needs, but also the metabolic requirements and appetites of mutans streptococci in their critical role of creating the necessary bonding substances for stable biofilm upon existing or exfoliated calculus layered upon tooth enamel. It is not hard to see, in this interpretation, that our ancient ancestors, who, so abundantly lived in tune with the natural world, would have had no reason to contemplate such a thing as "brushing one's teeth."

Finally, let's also not forget that the lesions on our teeth — and mass lost to decay — is something we are going to have to deal with. But taking a vital lesson from our ancient ancestors, the more we naturalize our diet, the less this will be an issue. The main argument against tartar formation from the dental industry is that it seals in acidogenic bacteria, and the only way at them is to destroy them in their cavitation. This means grinding away tartar, enamel, and anything damaged along their pathway. But we know this battle tactic is problematic, because it is only a stop gap measure, and ignores the entire platform of healing mechanisms stretching from the tooth root to the formative tartar to our salivation glands. As long as diet is moved to the backburner, and removing dental plaque is standard procedure in the name of prevention, the door to dental disease is kept wide open.

On a much brighter note, the doors to natural dental care are also wide open. But it's up to us, you and I, to walk on through and do something about it, or remain strapped to the chair and suffer the fate our specie's doesn't really deserve.

Epilogue
Natural Dental Care of the Future

"I would unite with anybody to do right and with nobody to do wrong." — Frederick Douglass

By definition, the future of natural dental care (NDC) translates to a natural alliance with whatever endeavor best serves both the "I" (our bodies) and "us" (our oral microbiota), because that can only serve strong, healthy teeth. The evidence is abundant, as a species, we have failed to do this as a modern incarnation of our ancient Paleolithic ancestors with great teeth. The industrialization of our food chain has clamped our troubled teeth in braces, and it is our mission to break free and do something about it. There's more we can do beyond the interventions described in previous chapters. And here are some of my ideas.

Educate and advocate

Don't keep the information in this book to yourselves. Share it with others, even your dentist and their staff if you have more courage than me! Challenge people to think about how and why tooth decay occurs. Counter the 3 Tooth Truths mythology with NDC facts. Explain that nature endowed the teeth with healing mechanisms from root to tartar. And that we can facilitate and support these "healing powers of nature" through a reasonably natural diet, exercising, and nutritional dental pastes that balance our microbiota and aid them in building strong tartar structure. For sure you will be met with snicker and anger when you challenge them. But, later, their defenses will begin to droop from confusion and hints of insecurity as they are left alone to contemplate the undeniable ravages that have taken place in their own mouths. Take refuge in the seminal wisdom of the German philosopher, Arthur Schopenhauer:

> All truth passes through three stages:
> First, it is ridiculed.
> Second, it is violently opposed.
> Third, it is accepted as being self-evident.

Support the Fluoride Action Network (FluorideAlert.org)

Fluoride is the #1 enemy of NDC. Visit the Fluoride Action Network[1] to learn why if you don't believe what this book has brought before you. At their

[1]https://fluoridealert.org

website you will see the findings of researchers that is suppressed by governments and their corrupt health agencies; which communities have fluoridated their water delivery systems and those that have refused; ongoing litigations; and more information that I can provide here. I leave this discussion with this quote posted on their website:

> "The political profluoridation stance has evolved into a dogmatic, authoritarian, essentially antiscientific posture, one that discourages open debate of scientific issues." - Dr. Edward Groth, Senior Scientist, Consumers Union, 1991.

Lobby our food chain industrialists

What this means is providing them with ideas that can improve their products. *Safer products.* The organics food industry is paving the way, in part. And the industrialists are watching and listening. They clearly want part of the action. That's good. What they all can do, and here I'm thinking of a concerted effort, is to begin dealing with the free sugar "addiction." Addicts must have their fix, and so we have a problem. How to undo it to safe levels. My recommendation is that the addiction can be reined in by the gradual, every so slightly, systematic removal of free sugars (natural and artificial) from processed foods. Like a person who gradually cuts back from their nicotine habit by smoking fewer and fewer cigarettes until they make the commitment to quit. To suddenly just jump over finish line just isn't workable for most people to kick their habits. Including sugar addiction.

Removing free sugars, along with unnecessary preservatives and other chemicals that wreak havoc on our microbiota, is something that can be done. And I see that some of the big industrialists are already doing this. Writing letters/emails of encouragement is probably not a bad idea. In fact, I don't think it's a waste of time at all. Even Walmart is now stocking nearly every aisle with non-GMO and organic foods. Although their buyers are ensconced at Wal-Mart's headquarters in Northwest Arkansas, they aren't a pack of suspicious hillbillies set in their ways. Upper management is yuppified, smart, hungry for profits, and progressive to the extent their ever-expanding grass roots don't bail on them.

Lobby your dentist

Since you're probably still in the chair, I'll presume that you still have lines of communication open with your dentist. Take this opportunity to ask them about non-invasive dental procedures actually performed by dentists that don't require drilling away decay and scraping off tartar, or using fluoride treatments. Ask them also if they can augment their staff with a certified herbalist and organic foods nutritionist.

S. Mutan vaccinations

Research is ongoing in this direction. I oppose this as *S. mutans* is a natural and significant player in creating biofilm that is formative to the natural tartar barrier. Diet and exercise as part of NDC favor boosting our immune systems.

Stem cell research

Stem cell research is ongoing "big time" relative to dental science, although you would never know that spending time in a dental office. A broad consortium of scientists involved are well aware of human suffering due to dental tissue disease, including trauma to supportive musculoskelature.[1] The core of their research mission involves introducing newly fabricated biomaterials into the body that serve as transporters for cells and control factors (chemicals) that coax extracellular tissue regeneration. The suggestion is that this relatively noninvasive intervention will recruit and stimulate host (stem) cells responsible for facilitating regeneration.

Peptide hydrogel therapy

This is a substance that when applied to micro-cavities in the tooth's enamel layer, forms a gel that combines with naturally deposited calcium from saliva. In this way, according to researchers, "It works by mimicking the protein scaffold around which tooth enamel naturally assembles itself as our teeth grow."[2] Within months, they report, cavitation limited to the enamel is repaired. While they state that "fluoride helps to re-grow enamel," it's not clear from their studies if the same thing happens without it. Further, the reparative regeneration is not replicated in the dentin, hence, therapy is limited, although important. The good news is that the procedure is painless and no drilling is required! A related article adds, "one of the main reasons people don't visit their dentists is fear. The researchers hope that by offering a pain-free solution to repairing teeth there will no longer be a reason to avoid the dentist's chair." Hold on, don't start heading back to the chair, your dentist probably isn't offering this service yet. Just tell them about it.

---

[1] https://mooneylab.seas.harvard.edu/musculoskeletal-tissue-engineering; also significant: http://orion.bme.columbia.edu/lulab/research/dental.htm

[2] https://medicinehealth.leeds.ac.uk/dir-record/research-projects/788/filling-without-drilling

Electrically Accelerated and Enhanced Remineralization (EAER)

This appears to be another enamel based therapy that requires no drilling or injections. So says Professor Nigel Pitts, from King's College London's Dental Institute, said: "The way we treat teeth today is not ideal. When we repair a tooth by putting in a filling, that tooth enters a cycle of drilling and refilling as, ultimately, each 'repair' fails."[1] Don't we all know about that? According to Pitts, the cavitation is first prepared and then subjected to a light electric current (painless) that corrals calcium and phosphate minerals into the site, therein stimulating the tooth's own reparative mechanisms. I discussed this in Chapter 4 in terms of nature's healing mechanisms, so I think this is definitely an important direction that fuses perfectly with natural dental care.

Related articles of interest

"Is pine tree bark the key to healthier, stronger teeth?" (https://www.naturalnews. com/2017-04-08-is-pine-tree-bark-the-key-to-healthier-stronger-teeth.html)

"Body's Own Stem Cells Can Lead to Tooth Regeneration – The End of Dental Implants?" (https://www.periodontal.com/bodys-stem-cells-can-lead-tooth-regeneration-end-dental-implants/)

"Researchers use light to coax stem cells to repair teeth." (https://seas.harvard. edu/news/2014/05/researchers-use-light-coax-stem-cells-repair-teeth)

In conclusion . . .

So ends this second edition of *Guard Your Teeth*. It is clear that new science is now pushing the frontiers of modern dentistry away from the mechanically invasive procedures of the chair. Regrettably, the important work of this vast collaboration of scientists remains firmly in the shadow of the chair. But, from what I've learned of them from reading their abstracts and lectures, they are a resolute force for change. At the same time, contemporary herbalists are plying their equally important trade in the same direction. Their intersection is the natural homeland of "natural dental care." These are exciting times, and the push is on to end and replace the modern version of Neolithic bow drills with stimulated tissue regeneration at the molecular level based on nature's secreted healing mechanisms. Both herbalist and scientist are entering this biological matrix in their own ways. *Guard Your Teeth* provides opportunities for direct action that require no authority other than your own common sense. May the "healing powers of nature" be with you!

[1]https://www.theguardian.com/society/2014/jun/16/fillings-dentists-tooth-decay-treatment; https://www. kcl.ac.uk/news/new-research-identifies-potential-treatment-to-manage-effects-of-periodontitis

Unless otherwise indicated below, images in this book are the author's.

Cover

- Front & back, except of the author: alphaspirit © www.123rf.com

P. 2

- Volodymyr Gorban © www.123rf.com

P.3

- tigatelu © www.123rf.com
- studiostoks © www.123rf.com

P.5

- https://en.wikipedia.org/wiki/Bridge_(dentistry)#/media/File: Bridge_from_dental_porcelain.jpg
- https://en.wikipedia.org/wiki/Dentures#/media/File:Mr_M%27s_Complete_Denture2.jpg
- https://en.wikipedia.org/wiki/Dental_implant#/media/File: Single_crown_implant.jpg
- https://en.wikipedia.org/wiki/Crown_(tooth)#/media/File: Blausen_0863_ToothAnatomy_02.png
- https://en.wikipedia.org/wiki/Endodontic_therapy#/media/File: Blausen_0774_RootCanal.png

P. 8

- Jill Willis

P. 9

- estt © www.123rf.com

P. 14

- https://en.wikipedia.org/wiki/Pleistocene_megafauna#/media/File: Ice_age_fauna_of_northern_Spain_-_Mauricio_Ant%C3%B3n.jpg

P. 15

- https://en.wikipedia.org/wiki/Caveman#/media/File:Caveman_5.jpg
- https://en.wikipedia.org/wiki/Paleo-Indians#/media/File: Glyptodon_old_drawing.jpg
- https://en.wikipedia.org/wiki/British_Agricultural_Revolution#/media/File:Maler_der_Grabkammer_des_Sennudem_001.jpg
- https://en.wikipedia.org/wiki/Combine_harvester#/media/File:Combine-harvesting-corn.jpg

P. 18

- http://laoblogger.com/image-post/67713-bad-tooth-clipart-8.jpg.html
- lightwise © www.123rf.com
- https://en.wikipedia.org/wiki/Tooth_decay#/media/File: Dental_Caries_Cavity_2.JPG

P.19

- *Nutrition and Physical Degeneration: A Comparison of Primitive and Modern Diets and Their Effects.* Weston A. Price.

P. 28

- filipefrazao © www.123rf.com

P. 29

- Filipefrazao © www.123rf.com

P. 39

- vecton © www.123rf.com

P. 41

- "Enamel significance in operative dentistry/ certified fixed orthodontic courses." Indian Dental Academy (www.indiandentalacademy.com)

P. 42

- "Clinical Significance of Dental Anatomy, Histology, Physiology, and Occlusion." Lee W. Boushell and John R. Sturdevant.

P. 45

- alphaspirit © www.123rf.com

P. 48

- microgen © www.123rf.com

P. 50-51

- http://clipartix.com/wp-content/uploads/2016/04/Tooth-clipart-clipartion-com-2.png

P. 53

- https://en.wikipedia.org/wiki/Toothpaste#/media/File:Toothpasteonbrush.jpg

P. 56

- https://en.wikipedia.org/wiki/Mouthwash#/media/File:Listerine_products.jpg

P. 58

- https://commons.wikimedia.org/wiki/File: Blausen_0864_ToothDecay.svg

P. 61

- tinna2727 © www.123rf.com
- ayphoto © www.123rf.com
- ayphoto © www.123rf.com
- thamkc © www.123rf.com
- joytasa © www.123rf.com

P. 62

- https://en.wikipedia.org/wiki/Disclosing_tablets#/media/File: Plaque_Disclosing_Tablets.jpg

P. 67

- https://www.nidcr.nih.gov/OralHealth/OralHealthInformation/ChildrensOralHealth/ToothDecayProcess.htm
- theartofphoto © www.123rf.com

P. 76

- https://en.wikipedia.org/wiki/Methamphetamine#/media/File: Suspectedmethmouth09-19-05.jpg
- robeo /© www.123rf.com

P. 77

- robertprzybysz © www.123rf.com

P. 80

- https://en.wikipedia.org/wiki/The_Flintstones#/media/File: Fred_and_Wilma_Flintstone_advertising_cigarettes.jpg

P. 89

- https://en.wikipedia.org/wiki/Obesity#/media/File:Obesity_%26_BMI.png

P. 90

- https://www.flickr.com/photos/nottinghamvets/6241445782/in/photolist-avx3ob
- J. Jackson. *Laminitis: An Equine Plague*, p. 35.

P. 99

- Jill Willis

P. 106

- wisiel © www.123rf.com
- Tanaporn Phothikhet © www.123rf.com
- Vasily Pindyurin © www.123rf.com
- undrey © www.123rf.com
- undrey © www.123rf.com
- Viacheslav Iakobchuk © www.123rf.com

P. 108

- Aleksandr Davydov © www.123rf.com
- Jill Willis

P. 109

- Jack LaLanne 1961. Photographer: Cliff Riddle. Public Domain.
- Jack LaLanne. Photographer: Unknown. Public Domain.

P. 110

- Aleksander Kaczmarek © www.123rf.com
- Aleksander Kaczmarek © www.123rf.com
- Aleksander Kaczmarek © www.123rf.com
- Maridav © www.123rf.com

P. 112

- Antonio Diaz © www.123rf.com
- Antonio Diaz © www.123rf.com

P. 113

- Dolgachov © www.123rf.com
- Ariwasabi © www.123rf.com
- Sergejs Rahunoks © www.123rf.com

P. 114

- GeorgeStepanek
- yobro10 © www.123rf.com
- Vladimir Voronin © www.123rf.com

P. 115

- Ritu Jagya © www.123rf.com

P. 116

- Raquel Baranow : https://en.wikipedia.org/wiki/Physical_exercise#/media/File: Woman_running_barefoot_on_beach.jpg
- Ferli Achirulli © www.123rf.com

P. 119

- Roman Tiraspolsky © www.123rf.com

I've always been a maverick thinker and doer, never satisfied with life's limits as I've perceived them to be in the mainstream. For example, after leaving the U. S. Army in early 1970 with an honorable discharge, I joined other veterans in the antiwar movement in protest of the corporate "war for profits" in Vietnam and the average American's unwitting complicity. "No business as usual" was our mantra, and on many fronts the burgeoning protest movement confronted every institution across the country. As mounting numbers of dead and wounded were returned home, the entire nation began questioning and then demanding an end to the war. In 1975 President Nixon felt the hand of the movement and shut it down — the greatest military blunder in the history of the U. S. After that, we all went own separate ways.

My calling became "nature" and what we can learn as a species from our natural world, past and present. And this is what my books are about. Come back often as my library will continue to grow.

www.ingramcontent.com/pod-product-compliance
Lightning Source LLC
Chambersburg PA
CBHW080558030426
42336CB00019B/3242